欧盟烟草制品指令

欧洲议会和欧盟理事会
2014/40/EU指令

胡清源　侯宏卫　等◎译

科学出版社

北京

内 容 简 介

《欧盟烟草制品指令》的目的是统一欧盟各成员国的相关法律、法规和管理规定，以促进烟草及其相关产品的境内市场平稳运作，并作为高度保护公众尤其是青少年健康的基础，使欧盟更好地履行对世界卫生组织《烟草控制框架公约》的义务。主要涉及领域包括：①烟草制品的成分和释放物及其相关报告义务，如卷烟焦油、烟碱和一氧化碳的最大释放量；②烟草制品的标识和包装，如烟草制品单位烟包及任何外包装上的健康警示，以及适用于烟草制品的确保其符合本指令要求的可追溯性和防伪标志；③禁止口用烟草制品投入市场；④烟草制品跨境远程销售；⑤提交关于新型烟草制品通告的义务；⑥一些新型烟草制品投放市场及其标识，如电子烟和贮液容器以及抽吸型草本制品等。

本书会引起吸烟与健康、烟草化学和公共卫生学等诸多应用领域的科学家的兴趣，为客观评价烟草制品的管制和披露提供必要的参考。

图书在版编目(CIP)数据

欧盟烟草制品指令：欧洲议会和欧盟理事会2014/40/EU指令 / 胡清源等译. —北京：科学出版社，2015.6
ISBN 978-7-03-045130-9

Ⅰ. ①欧…　Ⅱ. ①胡…　Ⅲ. ①欧洲国家联盟－烟草制品－标准　Ⅳ. ①TS45-65

中国版本图书馆CIP数据核字(2015)第133730号

责任编辑：刘　冉 / 责任校对：韩　杨
责任印制：徐晓晨 / 封面设计：铭轩堂

科学出版社 出版
北京东黄城根北街16号
邮政编码：100717
http://www.sciencep.com

北京教图印刷有限公司 印刷

科学出版社发行 各地新华书店经销

*

2015年6月第　一　版　开本：890 × 1240　A5
2015年6月第一次印刷　印张：3　7/8
字数：120 000

定价：50.00元

（如有印装质量问题，我社负责调换）

译 者 名 单

胡清源　侯宏卫　陈　欢
韩书磊　刘　彤　付亚宁

I

（立法法案）

指令

欧洲议会和欧盟理事会 2014 /40/EU 指令

2014 年 4 月 3 日

关于统一各成员国有关烟草及其相关产品生产、描述和销售的法律、法规和管理规定，并废止 2001/37/EC 指令

（与欧洲经济区 (EEA) 相关的文本）

欧洲议会和欧盟理事会，

考虑到欧盟运行条约，特别是第 53（1），62 和 114 条，

考虑到欧盟委员会的建议，

立法法案的草案发送到各国议会后，

考虑到欧洲经济和社会委员会的意见 [1]，

考虑到各地区委员会的意见 [2]，

按照普通立法的程序运作 [3]，

鉴于：

---

(1) OJ C 327, 12.11.2013, p.65。

(2) OJ C 280, 27.9.2013, p.57。

(3) 2014 年 2 月 26 日欧洲议会提出（未在官方杂志上发表），2014 年 3 月 14 日由欧盟理事会决议。

（1）欧洲议会和欧盟理事会2001/37/EC指令[1]在欧盟层面规定了有关烟草制品的条例。为了体现科学、市场和国际发展，该指令需要有实质性的变化，因此其应该被废止，并以新的指令取代。

（2）在2005年和2007年关于2001/37/EC指令执行情况的报告中，欧盟委员会确定了有益于境内市场平稳运作的深入举措的相关领域。在2008年和2010年，新兴及新鉴定健康风险科学委员会（SCENIHR）向欧盟委员会提供了对无烟烟草制品和烟草添加剂的科学性建议。2010年，开展了广泛的利益相关者磋商，紧接着是有针对性的利益相关者磋商并伴随着外部磋商。整个过程涉及各成员国。欧洲议会和欧盟理事会多次要求欧盟委员会审查和更新2001/37/EC指令。

（3）在2001/ 37/EC指令所涵盖的某些领域，成员国在法律上或在实际中受到了阻碍，阻止其有效地修改立法以适应新发展。尤其在标识相关条例上，成员国还未被允许增加健康警示的尺寸，在独立包装（“单位烟包”）上改变警示的位置或更换关于焦油、烟碱和一氧化碳（TNCO）释放水平的误导性警示。

（4）在其他领域，在成员国关于烟草及其相关产品生产、描述和销售的法律、法规和管理规定之间，仍存在实质性的差异，阻碍了境内市场的平稳运作。鉴于科学、市场和国际发展，这些差异可能会逐渐增加。该情况也同样存在于电子烟及电子烟的填充容器（“贮液容器”），抽吸型草本制品，烟草制品的成分和释放物，以及标识和包装及烟草制品跨境远程销售等方面。

（5）这些障碍应该被消除，为此，应进一步统一有关烟草及其相关产品生产、描述和销售的条例。

（6）鉴于烟草及其相关产品的境内市场规模，越来越多的烟草

(1) 2001年6月5日欧洲议会和欧盟理事会关于统一各成员国有关烟草制品生产、描述和销售的法律、法规和管理规定的2001/37/EC指令(OJ L 194, 18.7.2001, p. 26)。

制品制造商将供应于整个欧盟的产品集中在欧盟内的少量工厂生产，以及由此产生的烟草及其相关产品的重大的跨境贸易，均要求在欧盟层面采取更强的立法行动，而不是在国家层面实现境内市场的平稳运作。

（7）为了履行在2003年5月发布的世界卫生组织（WHO）《烟草控制框架公约》（FCTC）以及欧盟及其成员国的相关规定，在欧盟层面的立法行动也是必要的。关于烟草制品成分的规定，烟草制品信息披露、包装和标识以及广告和烟草制品非法贸易的规定的FCTC条款，都是特别相关的。为了FCTC条款的执行，FCTC的缔约方，包括欧盟及其成员国，在各种会议中通过取得共识的方式采纳了一系列的指导原则。

（8）按照欧盟运行条约（TFEU）的第114（3）条，高度的健康保护应该作为立法提案的基础，尤其是基于科学事实的任何新的进展都应被考虑进去。烟草制品不是普通商品，从其对人体健康特别有害的角度来看，健康保护应给予高度重视，尤其应降低青少年吸烟率。

（9）有必要给出一些新的定义，以确保本指令可以被成员国一致应用。本指令要求不同类别的产品履行不同的义务，对于同时属于多个类别范畴的相关产品（如烟斗、手卷烟），应遵从更严格的义务要求。

（10）2001/37/EC指令设定了卷烟中焦油、烟碱和一氧化碳释放量的最高限量，该限量也应适用于从欧盟出口的卷烟。该最高限量及其方法仍然有效。

（11）为了检测卷烟的焦油、烟碱和一氧化碳释放量（以下简称“释放量水平”），应参考有关的国际认可的ISO标准。验证过程中应采用独立的实验室（包括国家实验室），以免受烟草行业的影响。成员国应能使用欧盟其他成员国的实验室。对于烟草制品的其他释放物，没有国际公认的标准或量化其最大水平的测试。应当鼓励在国际层面上为制定这样的标准或测试进行不懈的努力。

（12）关于设定最大的释放量水平，有必要在未来降低焦油、烟碱和一氧化碳的释放量水平，或者依据烟草制品其他释放物的毒性或致瘾性，对其设定最大释放量。

（13）为了执行监管任务，各成员国和欧盟委员会需要关于烟草制品成分和释放物的综合信息来评估其吸引力、致瘾性和毒性，以及与该产品消费相关的健康危害。为此，现有的关于成分和释放物的报告义务应加以扩大。扩大的报告义务应包括在优先清单中的添加剂，以评估其自身毒性、致瘾性，以及致癌性、致突变或生殖毒性（“CMR 特性”），包括在燃烧形式下的相关特性。对于中小企业，这种增加的报告义务的负担应限制在可能的范围内。该报告义务与欧盟确保高度保护人体健康的义务是一致的。

（14）目前使用不同报告格式的情况使得制造商和进口商很难履行其报告义务，并且使得各成员国和欧盟委员会从获取的信息中比较、分析并得出结论的工作显得繁重。因此，应该对成分和释放量报告设定一个通用的强制性格式。应确保公众获得产品信息的最大可能的透明度，同时确保采用适当的账户以保护烟草制品制造商的商业秘密。现有的成分报告系统也应加以考虑。

（15）对烟草制品的成分管制方法尚不统一，影响了境内市场的平稳运作，并对欧盟的货物自由流通产生了负面的影响。一些成员国已经通过立法或与行业制订有约束力的协议以允许或禁止某些成分。结果，有些成分在某些成员国被管制，而在其他成员国不被管制。对于卷烟滤嘴中的添加剂以及使烟草烟气着色的添加剂，各成员国也采取不同的管制方法。考虑到 FCTC 及其相关准则在整个欧盟的实施以及在欧盟以外的其他司法管辖区获得的经验，若不加以协调，境内市场平稳运作的障碍预计将在未来几年内增加。FCTC 关于烟草制品成分管制以及烟草制品成分披露管制实施指南中尤其要求禁止使用具有能量和活性或有着色功能，可以增加适口性，暗示烟草制品有健康益处的成分。

（16）考虑到烟草制品可能含有烟草之外的一种特征香味，以促进烟草初吸或影响其消费方式，进一步增加了对其进行差异管制的可能性。应避免在不同风味类型卷烟的检测中引入不对等的差异处理。但是，较高销量的具有特征香味的产品应在较长的时间内被逐步禁止，让消费者有充足的时间转而使用其他产品。

（17）特征香味烟草制品的禁令并非彻底阻止个别添加剂的使用，但它确实迫使制造商减少添加剂或者添加剂组合的使用程度，使得添加剂不再产生特征香味。用于制造烟草制品的必要的添加剂，例如用糖代替醇化过程中损失的糖，应该是允许的，只要不产生特征香味或增加产品的致瘾性、毒性或 CMR 特性。一个独立的欧洲顾问小组应协助制定相关决策。本指令的执行不应该导致不同烟草品种之间的差别对待，也不应该妨碍产品差异化。

（18）某些添加剂被用于产生以下印象：烟草制品对健康有益，可降低健康风险，提神醒脑，增强机体功能。这些添加剂，以及在未燃烧形式下具有 CMR 特性的添加剂均应被禁止，以确保在整个欧盟规则统一，并高度保护人体健康。增加致瘾性和毒性的添加剂也应该被禁止。

（19）考虑到本指令关注于青少年，对于卷烟和手卷烟之外的烟草制品，在其销售量或青少年消费模式没有情况实质性改变时，应允许其在成分相关要求方面的豁免。

（20）考虑到在欧盟一般禁止出售口用烟草制品，出于对口用烟草制品的成分监管责任，需要对该产品的具体特性和消费模式进行深入了解，应与瑞典的辅助性原则一致，根据奥地利、芬兰和瑞典加入的法令的第 151 条，该产品被允许在瑞典销售。

（21）与本指令的目的相符，即促进烟草和相关产品的境内市场的平稳运作，以作为高度健康保护的基础，特别是针对青少年，并与欧盟理事会 2003/54/EC 建议[1]相一致，即鼓励各成员国通过采取适当的措施，制定和执行年龄限制，防止向儿童和青少年销售该类产品。

（22）烟草制品标识的相关规定在不同国家之间仍然存在不一致，特别是在单位烟包表面或内部包含的含图片和文本的组合健康警示、戒烟服务信息和促销因素等方面。

（23）这样的不一致可能形成贸易壁垒并阻碍烟草制品的境内市场平稳运作，因此，应被消除。此外，也可能造成一些成员国的消费者比其他成员国的消费者能更好地了解烟草制品的健康风险。如果没有在欧盟层面采取进一步行动，现有的差距有可能在未来几年内增加。

（24）为了使欧盟层面的相关规定顺应国际发展，关于标识相关规定的修改也是必要的。例如，FCTC 关于烟草制品包装标识的实施准则中提倡在两个主显示区都出现大幅的图片警示，强制包含戒烟信息，且关于限制误导性信息有严格规定。对误导性信息的规定将补充在欧洲议会和欧盟理事会 2005/29/EC 指令[2]的关于误导消费者的商业惯例的一般禁令中。

成员国出于财政的目的在烟草制品的包装上使用纳税印花或国家识别标志，在某些情况下，必须更改这些印花和标识的位置，以便在主显示区的顶端显示组合健康警示，与本指令和 FCTC 的指导原则相一致。本指令变更的过渡期安排应该到位，以允许成员国出于财政的

---

(1) 2002 年 12 月 2 日欧盟理事会关于吸烟预防及提高烟草控制措施的 2003/54/EC 建议（OJ L 22， 25.1.2003，p.31）。

(2) 2005 年 5 月 11 日欧洲议会和欧盟理事会关于境内市场企业对消费者不公平商业行为的 2005/29/EC 指令，修订的欧盟理事会 84/450/EEC 指令、97/7/EC 指令、98/27/EC 指令及欧洲议会和欧盟理事会 2002/65/EC 指令，欧洲议会和欧盟理事会（EC）2006/2004 号法规（“不公平商业行为指令”）（OJ L 149,11.6.2005，p.22）。

目的在单位烟包的顶端保持纳税印花或国家识别标志一段时间。

（25）对标识的规定也应随新的科学证据而变更。例如，卷烟单位烟包上标识的焦油、烟碱和一氧化碳的释放量水平已被证明是有误导性的，使消费者误认为某种卷烟比其他卷烟危害低。也有证据表明，包含文字警语和相应彩色图片的大型组合健康警示比只包含文字的警示更有效。因此，组合健康警示应该在整个欧盟成为强制性的，并且覆盖单位烟包表面的重要和可见的部分。应为所有的健康警示设定最小尺寸，以确保它们的可视度和有效性。

（26）对于卷烟和手卷烟之外的抽吸型烟草制品，其消费群体主要是年长的消费者和小众消费群体，只要销售量或青少年的消费模式没有情况实质性改变，就应该有可能继续给予它某些标识要求的豁免权。那些其他烟草制品的标识应遵循特定的规则。无烟烟草制品的健康警示的可见性应得到保证。因此，健康警示应被放置在无烟烟草制品包装的两个主表面上。至于水烟，它往往被视为比传统的抽吸型烟草制品危害性小，全部的标识规定都应当被应用于其上以避免消费者被误导。

（27）烟草制品或其包装可能误导消费者，特别是青少年，使他们认为这些产品危害性不高。这是由于使用了特定词语或特征，如“低焦油”、“清淡”、“超清淡”、“柔和”、“天然”、“有机”、“无添加剂”、“无调味剂”或“超细”，或特定的名称、图片，以及比喻性的图形或其他标志。其他误导内容可能包括，但不限于，附件或其他材料，如不干胶标签、贴纸、涂层和套管或与烟草制品本身形状相关的材料。特定的包装和烟草制品也可能通过诸如女人味、男人味、优雅等暗示其在减肥、异性吸引力、社会地位、生活品质方面的优势，从而误导消费者。同样地，单个烟支的大小和外观也可能制造出低害性的印象来误导消费者。无论是烟草制品的单位烟包，还是它们的外包装，都不应包含印刷代金券、折扣优惠、免费分发信息、买二赠一或其他类似的通过提出经济优势而刺激消费者购买这些烟草制品的相关优惠。

（28）为了确保健康警示的完整性和可见性以最大限度地发挥其效能，应对健康警示在烟草制品单位烟包上的相关尺寸以及视觉效果进行规定，包括形状和打开方式。当单位烟包的形状为长方体时，圆形或斜切的边缘被认为是可以接受的，其提供的健康警示的覆盖面积应与单位烟包上该边缘外的其他面积相当。成员国对于单位烟包内卷烟的最低数量有不同规定。这些规定应进行统一，以确保产品的自由流通。

（29）有相当数量的非法产品，不符合 2001/37/EC 指令的要求而被投放市场，而且有迹象表明，其销量可能会增加。这些非法产品破坏了合法产品的自由流通和烟草控制立法提供的保护。此外，FCTC 要求欧盟打击非法烟草制品，包括非法进口到欧盟的产品，并作为欧盟全面烟草控制政策的一部分。因此，必须为烟草制品的单位烟包制定规定，标明一个特有标识符和防伪标志以便更好地记录它们的运输，实现该产品在整个欧盟的跟踪和追溯，及更好地监测和执行其对本指令的依从性。此外，应为防伪标志的引入制定规章，这将有助于验证烟草制品的真伪。

（30）应在欧盟层面开发一种可互操作的跟踪和追踪系统及防伪标志。在最初阶段只在卷烟和手卷烟方面使用跟踪和追踪系统及防伪标志。这将使得其他烟草制品制造商在该跟踪和追踪系统及防伪标志应用于其他产品前获得经验。

（31）为了保证跟踪和追踪系统的独立性和透明度，烟草制品制造商应与独立的第三方签订数据存储合约。欧盟委员会应批准这些独立第三方的适用性，并有独立的外部审计人员监督他们的活动。与跟踪和追踪系统相关的数据应该与其他公司的相关数据分开，由各成员国主管部门和欧盟委员会控制并可以在任何时间访问。

（32）欧盟理事会 89/622/EEC 指令[1]禁止成员国销售特定类型的口用烟草制品。2001/37/EC 指令重申了该禁令。奥地利、芬兰和瑞典加入的法令的第 151 条批准了瑞典对该禁令的豁免。应继续实施对口用烟草制品的销售禁令，以阻止在欧盟（除瑞典外）引入有致瘾性和对健康有不利影响的产品。对于其他非面向大众市场的无烟烟草制品，对包装标识的严格规定和与成分相关的规定被认为足以遏制其在传统使用之外的市场扩张。

（33）烟草制品的跨境远程销售会导致消费者获取不符合本指令的烟草制品。这也将增加青少年获取烟草制品的风险。其结果有可能会破坏烟草控制立法。因此，各成员国应被允许禁止跨境远程销售烟草。在跨境远程销售不禁止的地方，对从事这种销售的零售店实行通用的登记规则是适当的，以保证本指令的有效性。各成员国应按照欧盟条约（TEU）的第 4（3）条相互合作以促进本指令的实施，特别是对于烟草制品跨境远程销售的有关措施。

（34）所有烟草制品都有引起死亡、疾病和致残的潜力。因此，应对它们的生产、分配和消费进行监管。对新型烟草制品的发展的监管是很重要的。在不损害各成员国禁止或批准新型烟草制品的权力的情况下，制造商和进口商有义务提交该新型烟草制品的通告。

（35）为确保一个公平竞争的环境，按本指令定义属于烟草制品的新型烟草制品应符合本指令的要求。

（36）电子烟和贮液容器应受本指令管控，除非基于其描述或功能，它们受制于欧洲议会和欧盟理事会 2001/83/EC 指令[2]或欧盟

(1) 1989 年 11 月 13 日统一各成员国有关烟草制品标识和禁止销售特定类型口用烟草制品的法律、法规和管理规定的欧盟理事会 89/622/EEC 指令（OJ L 359,8.12.1989，p.1）。

(2) 2001 年 11 月 6 日欧洲议会和欧盟理事会关于人用药品的共同体法典的 2001/83/EC 指令。

理事会 93/42/EEC 指令[1]。对此类产品的立法和实践在成员国之间存在差异，包括对安全性的要求等。因此，需在欧盟层面采取行动，以促进境内市场的平稳运作。对此类烟草制品进行管制时应考虑高度保护公众健康。为了便于各成员国执行其监视和控制任务，应要求电子烟和贮液容器的制造商和进口商在产品投放市场之前，必须提交一份相关产品的通告。

（37）各成员国应确保电子烟和贮液容器符合本指令的要求。若相关产品的制造商不在欧盟境内，则该产品的进口商应承担关于该产品服从本指令的义务。

（38）本指令要求，对于含有烟碱的烟液，当其烟碱浓度不超过 20 mg/mL 时才允许投放市场。此浓度与相同时间内抽吸标准卷烟所能得到的烟碱剂量相当。为了限制烟碱相关的风险，应设置贮液容器、烟液池和烟弹的最大尺寸。

（39）本指令要求，只有烟碱传输量保持稳定水平的电子烟被允许投放市场。在正常使用条件下，出于对健康保护、安全和质量的目的（包括避免意外的高剂量摄入风险），保证烟碱传输量的稳定性是必要的。

（40）电子烟和贮液容器被儿童获得可能引起健康风险。因此，有必要确保此类产品是对儿童安全的，且可防止随意摆弄，包括通过设置标识，加固和打开机关等以确保儿童安全。

（41）考虑到烟碱是一种有毒物质，且有潜在的健康和安全风险，包括对于不打算使用含烟碱产品的人。因此，含有烟碱的烟液只能置于满足特定安全和质量要求的电子烟或贮液容器内投放市场。确保电子烟在使用和重新注液时不破裂或泄漏很重要。

---

(1) 1993 年 6 月 14 日欧盟理事会关于医疗器械的 93/42/EEC 指令（OJ L 169,12.7.1993, p.1）。

（42）这些产品的标识和包装应显示充分和适当的安全使用信息，以保护人体健康和安全，应进行适当的健康警示，而不应包含任何具有误导性的要素或特征。

（43）关于电子烟的广告和赞助在国家法律和实践之间存在的差距，阻碍了商品的自由流通和提供服务的自由，并制造了扭曲竞争的巨大风险。另外考虑到电子烟和贮液容器市场的不断增长，若不在欧盟层面采取进一步行动，这些差异很可能会在未来几年内增加。因此，有必要统一关于这些产品的具有跨境影响的广告和赞助的国家规定，从而高度保护人体健康。因为电子烟可以模仿吸烟行为并使其常规化，故而电子烟可以发展为烟碱成瘾的入门工具并最终发展为传统烟草的消费。为此，在电子烟和贮液容器的广告方面宜加以限制。

（44）为执行其监管任务，欧盟委员会和各成员国需要关于电子烟和贮液容器市场发展的综合性信息。为此，这些产品的制造商和进口商有报告销量、不同消费群体的偏好以及销售模式的义务。在考虑到保护商业秘密的前提下，应保证将这些信息提供给公众。

（45）为确保各成员国有适当的市场监督，制造商、进口商和分销商有必要通过一个适当的系统来监测和记录可疑的负面影响，并将这些影响上报主管部门，以便采取适当的行动。有必要提供一个保护条款，允许成员国采取行动来应对严重的公众健康风险。

（46）在电子烟作为一个新兴市场的背景下，尽管符合本指令要求，投放市场的特定电子烟或贮液容器，或一类电子烟或贮液容器，仍可能对人体健康造成无法预料的风险。因此建议提供一个程序以应对这项风险，包括成员国采取临时的适当措施的可能性。这种临时的适当措施可能涉及禁止特定电子烟或贮液容器，或者一类电子烟或贮液容器投放市场。在这种情况下，欧盟委员会应有权采用授权法案，禁止将特定电子烟或贮液容器，或者一类电子烟或贮液容器投放市场。当至少三个成员国基于充分正当理由禁止有关产品时，欧盟委员会应有权将该禁令扩大至所有成员国，以确保符合本指令而无相同健康风险的产品在境内市场的平稳运作。欧盟委员会应在2016年5月20日*前提交可填充电子烟相关潜在风险的报告。

（47）本指令不统一电子烟或贮液容器的各个方面。例如，对香味成分管制的责任仍然在各成员国。这对于成员国考虑允许加香产品投放市场是有用的。在这样做时，他们应该考虑到这类产品对青少年和不吸烟者的潜在吸引力。任何禁止这种加香产品的规定都需要有正当理由并依照欧洲议会和欧盟理事会98/34/EC指令(1)提交通告。

（48）此外，本指令不统一无烟环境的规则，或者国内销售计划或国内广告，又或者品牌延伸，也不设定电子烟或贮液容器的年龄限制。在任何情况下，这些产品的描述和广告不应该促进烟草消费或引起烟草制品的混乱。鼓励各成员国在管辖权范围内自由管控这些事项。

（49）对抽吸型草本制品的管制在不同成员国间存在差异，这些产品通常被认为是无害或危害较小的，尽管其燃烧同样引起健康风险。在许多情况下，消费者并不知道这些产品的成分。为了确保境内市场的平稳运作并增加给消费者的信息，应在欧盟层面出台这些产品的通用标签规则和成分报告。

（50）为了确保本指令执行的一致性，应授予欧盟委员会在以下方面的执行权力：制定和更新添加剂的优先清单以便于扩大报告的范围，制定和更新成分披露报告的格式以便于相应信息的传播，确定烟草制品是否具有特征香味或增加了毒性、致瘾性或CMR特性，确定烟草制品是否具有特征香味的方法学，为确定烟草制品特征香

---

*　本指令生效2年后。——译注

(1)　1998年6月22日欧洲议会和欧盟理事会关于在技术标准和法规领域提供信息程序以及关于信息社会服务规则的98/34/EC指令（OJ L 204,21.7.1998，p.37）。

味建立和运行独立顾问小组的程序，手卷烟烟袋上的健康警示的精确位置，组合健康警示的布局、设计和形状的技术规范，建立和运行跟踪和追踪系统的技术标准，确保特有标识符和防伪标志系统的兼容性，以及为电子烟、贮液容器和此类产品更换烟液原理的技术标准建立统一的通告格式。那些执行权力应当参照欧洲议会和欧盟理事会（EU）182/2011 号法规[(1)]执行。

（51）为了确保本指令是完全可操作的，以及为了使其适应烟草制造、消费和管制领域的科学、技术和国际发展，依据 TFEU 第 290 条，修改法案的权力应赋予欧盟委员会，修改的内容涉及批准和改变最大释放量水平及释放量检测方法，为引起特征香味或增加毒性或致瘾性的添加剂设定最高限量，收回赋予卷烟和手卷烟之外的烟草制品的特定豁免权，更改健康警示，建立和变更图片库，定义用于跟踪和追踪系统的数据存储合约的关键要素，将成员国通过的关于特定电子烟或贮液容器或者一类电子烟或贮液容器的检测方法扩大至整个欧盟范围。在准备工作期间，欧盟委员会进行适当的磋商，包括专家级的磋商，是十分重要的。欧盟委员会在准备和制定授权法案时，应确保同时、实时和适时地将有关文件提交给欧洲议会和欧盟理事会。

（52）欧盟委员会应监测本指令实施及其影响的发展，并在 2021 年 5 月 21 日 * 前及以后的必要时候提交报告，以评估本指令的修订是否是必要的。报告应包括本指令监管外的烟草制品单位烟包表面的信息，新型烟草制品的市场发展，意味着情况实质性改变的市场发展，关于超细卷烟、水烟、电子烟和贮液容器的市场发展及消费者认知。

(1) 2011 年 2 月 16 日欧洲议会和欧盟理事会关于制定欧盟委员会行使授予的执行权力的规则和成员国控制机制的一般原则的 (EC)182/2011 号法规（OJ L 55, 28.2.2011, p.13）。

* 本指令生效 7 年后。——译注

欧盟委员会应准备一份关于欧洲范围内开展烟草制品成分管制的可行性、益处及影响的报告，包括在欧盟层面制定一个可用于或存在于又或添加到烟草制品的成分清单（即所谓的“肯定列表”）的可行性和益处。在报告准备过程中，欧盟委员会应重点评价关于成分毒性和致瘾性效果的有效科学证据。

（53）符合本指令要求的烟草及其相关产品应当从商品的自由流通中获益。然而，鉴于本指令的统一程度不同，出于保护公众健康的目的，各成员国在一定的条件下有权在某些方面施加进一步的要求，例如，除了本指令对健康警示设定的基础统一规则之外的烟草制品的呈现和包装（如颜色）等。因此，各成员国可以出台规定以进一步规范烟草制品的包装，但这些规定须与 TFEU 和 WTO 的义务相兼容，并且不影响本指令的全面执行。

（54）此外，为了顾及未来市场可能的发展，成员国也应被允许禁止某些类别的烟草及其相关产品，只要其出于成员国具体情况的考虑，以及该规定出于保护公众健康的正当理由，同时兼顾本指令达到的高水平保护。各成员国应将更严格的国家条款通告欧盟委员会。

（55）对于投放国内市场的所有产品，成员国可针对本指令管制外的其他方面自由维持或制定国家法律，只要该法律与 TFEU 兼容且不妨碍本指令的全面执行。因此，在上述条件下，成员国可以管制或禁止用于烟草制品（包括水烟）和抽吸型草本制品的相关装置，以及管制或禁止外观类似一类烟草或其相关产品的产品。根据 98/34/EC 指令要求，在国家技术管制之前要预先通告。

（56）各成员国应确保个人数据仅在依照欧洲议会和欧盟理事会 95/46/EC 指令[(1)]的规则和保护措施的情况下被处理。

(1) 1995 年 10 月 24 日欧洲议会和欧盟理事会关于个人数据处理和自由流通的保护的 95/46/EC 指令（OJ L 281，23.11.1995，p.31）。

（57）本指令不妨碍欧盟关于转基因作物使用和标识的监管法律。

（58）与 2011 年 9 月 28 日各成员国和欧盟委员会关于说明文件的联合政治宣言[1]相一致，在合理的情况下，成员国承诺在关于改变检测方法的通告中附带一份或几份文件，来解释指令中相关部件与国内替换仪器对应部件的关系。本指令立法者认为此类文件的传播是合理的。

（59）本指令有义务遵守欧盟基本权利宪章中记载的基本权利和法律原则。本指令影响到一些基本权利。因此必须确保烟草及其相关产品的制造商、进口商和分销商所履行的义务，不仅要保证高水平的健康和消费者保护，而且要保护所有其他基本权利，且有助于境内市场的平稳运作。本指令的执行应遵守欧盟法律和相关的国际义务。

（60）因为本指令的目标，即统一各成员国有关烟草及其相关产品生产、描述和销售的法律、法规和管理规定，不能在各成员国充分实现，但由于其规模和影响，可以更好地在欧盟层面实现，欧盟可以按照 TEU 第 5 条中规定的辅助性原则采取相关措施。依据该条款中设置的比例原则，本指令没有超出为实现这些目标所必需的范畴。

本法令规定：

第一部分　通　　则

第一节　主　　题

本指令的目的是统一各成员国的相关法律、法规和管理规定，以促进烟草及其相关产品的境内市场平稳运作，并作为高度保护公众尤其是青少年健康的基础，使欧盟更好地履行对世界卫生组织《烟

(1)　OJ C 369,17.12.2011，p.14。

草控制框架公约》的义务。主要涉及领域包括：

（1）烟草制品的成分和释放物及其相关报告义务，如卷烟焦油、烟碱和一氧化碳的最大释放量；

（2）烟草制品的标识和包装，如烟草制品单位烟包及任何外包装上的健康警示，以及适用于烟草制品的确保其符合本指令要求的可追溯性和防伪标志；

（3）禁止口用烟草制品投放市场；

（4）烟草制品跨境远程销售；

（5）提交关于新型烟草制品的通告的义务；

（6）一些新型烟草制品投放市场及其标识，如电子烟和贮液容器以及抽吸型草本制品等。

第二节　定　　义

为本指令叙述方便起见，特给出以下定义：

（1）“烟草”是指烟叶或者其他天然加工或者未加工的烟草作物，包括膨胀和再造烟草。

（2）“烟斗烟草”是指通过燃烧过程消耗的专用于烟斗使用的烟草。

（3）“手卷烟”是指可供消费者或者零售商自己制造卷烟来使用的烟草。

（4）“烟草制品”是指可用于消费的、全部或者部分由烟草组成的产品，不论是否是转基因烟草。

（5）“无烟烟草制品”是指不涉及燃烧过程的烟草制品，包括嚼烟、鼻烟和口用烟草制品。

（6）“嚼烟”是指专门用于咀嚼的无烟烟草制品。

（7）“鼻烟”是指可以通过鼻子进行消费的无烟烟草制品。

（8）“口用烟草制品”是指除吸入和咀嚼之外的供口部使用的烟

草制品，其全部或部分由烟草制成，以粉末、颗粒或者多种状态的组合形式存在，特别是存在于药囊部分或多孔囊中。

（9）“抽吸型烟草制品”是指无烟烟草制品以外的其他烟草制品。

（10）“卷烟”是指通过燃烧过程来消费的卷制的烟草制品，其详细定义可参见欧盟理事会 2011/64/EU 指令[(1)]第 3(1) 条。

（11）“雪茄”是指通过燃烧过程来消费的卷制的烟草制品，其详细定义可参见欧盟理事会 2011/64/EU 指令第 4(1) 条。

（12）“小雪茄”是指一种小型的雪茄，其详细定义可参见欧盟理事会 2007/74/EC 指令[(2)]第 8(1) 条。

（13）“水烟”是指可以通过水烟筒来消费的烟草制品。在本指令中，水烟被视为抽吸型烟草制品。如果一种烟草制品既能通过水烟筒抽吸又能视为手卷烟，则将其视为手卷烟。

（14）“新型烟草制品”是指同时满足以下两个条件的烟草制品：

（a）不属于传统卷烟、手卷烟、烟斗烟草、水烟、雪茄、小雪茄、嚼烟、鼻烟或口用烟草制品范畴；

（b）在 2014 年 5 月 19 日 * 之后投放市场。

（15）“抽吸型草本制品”是指由植物、草药或水果制成，不含烟草，可通过燃烧过程来消费的产品。

（16）“电子烟”是指通过嘴端或产品的任何其他部分产生含烟碱的气溶胶，以供消费的产品，包括烟弹、烟液池和不含烟弹或烟液池的装置。电子烟可以是一次性的，也可以通过向贮液容器或烟液池重新填充烟液，或者更换一次性烟弹而重复使用。

(1) 2011 年 6 月 21 日欧盟理事会关于加工烟草制品适用的消费税结构和税率的 2011/64/EU 指令（OJ L 176,5.7.2011，p.24）。

(2) 2007 年 12 月 20 日欧盟理事会关于由第三国的旅客携带入关的物品可免征增值税和消费税的 2007/74/EC 指令 (OJ L 346，29.12.2007，p.6)。

* 本指令生效日期。——译注

（17）“贮液容器”是指一个装填含有烟碱的烟液的容器，用于重新填装电子烟。

（18）“成分”是指烟草、添加剂以及存在于最终烟草制品或相关产品中的任何物质或元素，包括卷烟纸、滤嘴、油墨、胶囊和胶黏剂。

（19）“烟碱”是指烟碱类生物碱。

（20）“焦油”是指不包含水和烟碱的烟气冷凝物。

（21）“释放物”是指烟草或其相关产品在有意消费过程中释放出来的物质，如卷烟烟气中的物质，或使用无烟烟草制品过程中释放的物质。

（22）“最高限量”或“最大释放量”是指烟草制品中某物质的最高含量或释放量，以毫克计，该数值可以为0。

（23）“添加剂”是指向烟草制品、单位烟包或任何外包装中添加的非烟草的其他物质。

（24）“调味剂”是指能影响香味和/或吃味的添加剂。

（25）“特征香味”是指来源于一种添加剂或多种添加剂组合产生的，不同于烟草的一种显著的香味或吃味，在烟草制品消费前或消费过程中均有显著特征，包括但不限于水果、香料、药草、酒精、糖果、薄荷或香草。

（26）“致瘾性”是指一种物质能够引起成瘾的药理学功效，会影响一个人控制其行为能力的状态，通常是通过逐渐奖赏或减免脱瘾症状方式实现，或两者共同实现。

（27）“毒性”是指某物质能够对人的机体产生不利影响的程度，包括随时间变化的影响，通常是通过反复或连续使用或暴露来实现。

（28）“情况实质性改变”是指按产品类型划分的某类烟草制品的销售量在至少五个成员国的增长都超过10%（根据第五节第6条的计算方法），或在不少于五个成员国内，以各自产品类型划分的某类烟草制品在25岁以下消费群体的流行度增长5个百分点以上（基于 2012 年 5 月的欧盟民意调查专题 385 报告或类似的流行性调查），但是，如果在零售水平上按烟草类型计算的销售量增长不超过欧盟层面烟草制品总销售量的 2.5%，则认为没有发生情况实质性改变。

（29）“外包装”是指投放市场的烟草或其相关产品的任何包装，包括单位烟包或单位烟包的集合，透明封套不视为外包装。

（30）“单位烟包”是指投放市场的烟草或其相关产品的最小独立包装。

（31）“烟袋”是指手卷烟的单位烟包，要么是一个长方形的口袋，在开口处有封条，要么是一个可长期使用的烟袋。

（32）“健康警示”是指对产品消费后可能引起的人体健康的不良影响或其他不良后果的警告，包括文字警语、组合健康警示、通用警语和信息，正如为本指令所提供的。

（33）“组合健康警示”是指由文字警语和相应图片或图示组合而成的健康警示，正如为本指令所提供的。

（34）“跨境远程销售”是指远距离销售给如下情况的消费者：在通过零售商（在成员国或其他国家设立）订购产品时，消费者居住于零售商所在国家之外的成员国。当零售商满足以下条件时，被视为在成员国设立：

（a）对于自然人，如果他或她在成员国里有自己的经营场所；

（b）其他情况，如果零售商有其法定所在地、业务管理中心或经营场所，包括在该成员国的一个分支、代理或任何其他机构。

（35）“消费者”是指出于贸易、商业、制作或职业以外目的的自然人。

（36）“年龄认证系统”是指根据国家要求自动确认消费者年龄的计算机系统。

（37）“制造商”是指生产产品或者设计、制造，并以其名字或商标销售产品的个人或法人。

（38）“烟草或其相关产品的进口”是指烟草或其相关产品进入欧盟领域，无论是否按海关监督程序或安排处置。

（39）“烟草或其相关产品的进口商”是指已经进入欧盟领域的烟草或其相关产品的拥有者，或者有权处理这些产品的人。

（40）“投放市场”是指不管烟草或其相关产品的生产地，通过免费或收费方式使其能够被欧盟的消费者购买到，包括远距离销售；对于跨境远程销售，产品被视为投放到消费者所在成员国的市场。

（41）“零售商”是指销售烟草制品的商店，也包括自然人。

第二部分　烟 草 制 品

第一章　成分和释放物

第三节　焦油、烟碱、一氧化碳和其他物质的最大释放量

1. 各成员国投放市场或者制造的卷烟的最大释放量（“最高限量”）应该不超过：

（a）焦油：每支卷烟 10 mg；

（b）烟碱：每支卷烟 1 mg；

（c）一氧化碳：每支卷烟 10 mg。

2. 欧盟委员会有权根据第二十七节采用授权法案，降低本节第 1 条所规定的最大释放量，但必须基于国际普遍认可的标准。

3. 各成员国应向欧盟委员会通告他们卷烟产品中不同于本节第 1 条中设定的最大释放量，以及卷烟之外烟草制品的最大释放量。

4. 欧盟委员会应根据第二十七节采用授权法案，将由 FCTC 各缔约方或由 WHO 通过的不同于本节第 1 条中设定的卷烟最大释放量标准以及卷烟之外烟草制品的最大释放量标准整合进欧盟法律。

第四节　检测方法

1. 卷烟焦油、烟碱和一氧化碳的释放量应分别参照 ISO 4387 标准、ISO 10315 标准和 ISO 8454 标准检测。

焦油、烟碱和一氧化碳检测的准确度必须符合 ISO 8243 标准的规定。

2. 本节第 1 条提到的检测应由成员国主管部门批准或监控的实验室进行验证。

上述实验室不能直接或间接由烟草公司拥有或者控制。

各成员国应提供给欧盟委员会一个获批准实验室的清单，明确其获得批准的标准和采用的监控方法，并在作出任何变更的时候及时更新该清单。欧盟委员会应将这些获批准实验室的清单向公众公开。

3. 如果基于科学技术发展或国际普遍认可的标准认为有必要的话，欧盟委员会有权根据第二十七节采用授权法案，修改焦油、烟碱和一氧化碳的检测方法。

4. 各成员国应向欧盟委员会通告他们所使用的本节第 3 条提到的释放物之外的卷烟释放物检测方法，以及卷烟之外烟草制品释放物的检测方法。

5. 欧盟委员会应根据第二十七节采用授权法案，将由 FCTC 各缔约方或由 WHO 通过的标准检测方法整合进欧盟法律。

6. 成员国可以向烟草制品的制造商和进口商收取一定比例的费用，用于本节第 1 条中相关检测的验证。

第五节　成分和释放物的报告

1. 各成员国应当要求烟草制品的制造商和进口商按品牌名称和产品类型向其主管部门提交如下信息：

（a）烟草制品生产过程中使用的所有成分及其含量清单，按各成分在烟草制品中的含量降序排列；

（b）本指令第三节第 1 条和第 4 条提到的释放量；

（c）如果可以的话，提供其他释放物信息及其含量水平。

对于已经投放市场的烟草制品，以上信息应在 2016 年 11 月 20 日 * 前提供。

如果烟草制品中某一成分被修改，从而影响按本节提供的相关信息，制造商或进口商应当告知相关成员国的主管部门。

本节要求的相关信息应当在新的或被改良的烟草制品投放市场之前提交。

2. 本节第 1 条（a）中提到的成分清单应附相关说明，阐明清单中的每一种成分在烟草制品中使用的原因。该清单还应当指出各成分的状态，包括是否按照欧洲议会和欧盟理事会（EC）1907/2006 号法规 [1] 登记，以及在欧洲议会和欧盟理事会（EC）1272/2008 号法规 [2] 中的分类。

3. 本节第 1 条（a）中提到的清单还应附相关毒理学数据，包括清单中的每一种成分在燃烧以及未燃烧状态下的相对毒性，可以的话，应特别给出其对消费者健康的影响尤其是致瘾性作用。

此外，对于卷烟和手卷烟，制造商和进口商应当提交一份技术

---

* 本指令生效 30 个月后。——译注

(1) 2006 年 12 月 18 日欧洲议会和欧盟理事会关于化学品登记、评估、授权和限制（REACH）的（EC）1907/2006 号法规，同时建立了欧洲化学品管理局，修订了 1999/45/EC 指令，废除了欧盟理事会（EEC）793/93 号法规、欧盟委员会（EC）1488/94 号法规和欧盟理事会 76/769/EEC 指令以及欧盟委员会 91/155/EEC 指令、93/67/EEC 指令、93/105/EC 指令和 2000/21/EC 指令 (OJ L 396, 30.12.2006, p. 1)。

(2) 2008 年 12 月 16 日欧洲议会和欧盟理事会关于物质和混合物分类、标识与包装的（EC）1272/2008 号法规，同时修订了 67/548/EEC 指令，废止了 1999/45/EC 指令，并修订了（EC）1907/2006 号法规（OJ L 353,31.12.2008，p.1）。

报告，对其中使用的添加剂及其性状进行总体描述。

除焦油、烟碱和一氧化碳以及第四节第 4 条提到的释放物外，制造商和进口商应当提供所使用的释放物检测方法。成员国应当要求生产商或进口商依据其主管部门颁布的法令，对烟草制品中成分的危害性进行评估，特别是其致瘾性和毒性。

4. 成员国应确保根据本节第 1 条及第六节第 1 条提交的信息在互联网上公开。在信息公开的同时，成员国还应考虑到充分保护商业机密的需要。成员国应当要求制造商和进口商在按照本节第 1 条及第六节第 1 条提交信息时，明确哪些信息被认为是商业机密。

5. 欧盟委员会应当通过执行法案，颁布法律，如果需要的话，改变本节第 1 条及第六节第 1 条提到的信息的提交形式和公开形式。这些执行法案应符合第二十五节第 2 条的审查程序。

6. 成员国应当要求制造商和进口商提交他们可获得的内部和外部市场调查，包括青少年和当前吸烟人群在内的不同消费群体的偏好研究，烟草成分及释放物研究，以及在新产品上市时所进行的任何市场调查。成员国还应当要求制造商和进口商报告每种品牌和产品类型的销售量，以条或者千克计。从 2015 年 1 月 1 日 * 开始，每个成员国以年销售量进行报告。各成员国应提供其他任何可获得的销售数据。

7. 本节及第六节要求成员国提供的所有数据和信息应以电子文件的形式提交。各成员国应当保存所提交的电子信息，并确保欧盟委员会及其他成员国出于适用本指令的目的能够合理使用这些信息。各成员国和欧盟委员会应当确保其中的商业机密及其他隐私信息的保密性。

8. 成员国可以向制造商和进口商收取一定比例的费用，用于接

---

* 本指令生效后第一个年度的开始。——译注

收、储存、处理、分析及公开本节所要求提交的信息。

第六节　添加剂的优先清单及扩大的报告义务

1. 除了第五节中规定的报告义务，应将报告义务加以扩大，包含被优先清单收录在内的卷烟及手卷烟的添加剂。欧盟委员会应当通过执行法案来建立以及更新添加剂的优先清单。这一清单应当包括以下添加剂：

（a）初始指征、调查研究或其他现有司法法规管控中表明具有本节第 2 条中（a）~（d）所述特性中至少一项的添加剂；以及

（b）依据第五节第 1~3 条报告的成分重量或数量，最常使用的添加剂。

这些执行法案应符合第二十五节第 2 条的审查程序。第一份添加剂清单应在 2016 年 5 月 20 日 * 前采用，而且应当包含至少 15 种添加剂。

2. 各成员国应当要求含有本节第 1 条优先清单中添加剂的卷烟及手卷烟的制造商和进口商对每一种添加剂进行全面的研究，考察其是否：

（a）对烟草制品的毒性或致瘾性有贡献，不论此贡献是否可将烟草制品的毒性或致瘾性增加到显著或可测量水平；

（b）会引起特征香味；

（c）能够促进吸入或烟碱摄入；或

（d）导致产生具有 CMR 特性的物质，不论其数量是否达到可将烟草制品的 CMR 特性增加到显著或可测量水平。

3. 上述研究应当考虑到烟草制品的有意使用，并特别考察燃烧

* 本指令生效 2 年后。——译注

过程中添加剂的释放问题。研究还应考察添加剂与烟草制品中其他成分的相互作用。在类似烟草制品中使用同一种添加剂的制造商或进口商应当开展联合研究。

4. 制造商或进口商应当报告上述研究的结果。报告应包含概要，关于添加剂的科学文献的全面综述，以及有关添加剂所产生影响的内部数据的总结。

制造商或进口商应当向欧盟委员会提交上述报告，同时向含有本节第 1 条优先清单中添加剂的烟草制品已投放市场 18 个月以上的成员国的主管部门提交报告副本。欧盟委员会和相关成员国也可以要求制造商或进口商提供有关添加剂的补充信息。这些补充信息应当作为报告的一部分。

欧盟委员会和相关成员国可以要求独立的科研机构对报告进行同行评议，特别是对其全面性、方法学及结论的审查。反馈的信息有助于欧盟委员会和各成员国根据第七节进行决策。各成员国和欧盟委员会可以向烟草制品的制造商和进口商收取一定比例的费用，用于上述同行评议。

5. 如果有其他制造商或进口商提供添加剂的相关信息，则依据欧盟委员会 2003/361/EC 建议[(1)]定义的中小型企业应当被豁免本节规定的相应义务。

第七节　成分管制

1. 成员国应当禁止具有特征香味的烟草制品投放市场。

成员国不能禁止在烟草制品生产过程中使用必要添加剂，例如加入糖以弥补在醇化过程中糖的损失，前提是这些添加剂的使用不

(1)　2003 年 5 月 6 日欧盟委员会关于微型、小型和中型企业定义的 2003/361/EC 建议 (OJ L 124, 20.5.2003, p. 36)。

会导致产品产生特殊香味，也不会导致烟草制品的致瘾性、毒性以及 CMR 特征增加到显著或可测量水平。

成员国应将依据本条采取的相应措施通告欧盟委员会。

2. 欧盟委员会应当在成员国的要求下或是自发地通过执行法案来确定某烟草制品是否属于本节第 1 条所指的范围。这些执行法案应符合第二十五节第 2 条的审查程序。

3. 欧盟委员会应当通过执行法案，为确定烟草制品是否属于本节第 1 条所指范围的程序颁布统一规定。这些执行法案应符合第二十五节第 2 条的审查程序。

4. 应当在欧盟层面成立一个独立的顾问小组。各成员国和欧盟委员会在依据本节第 1 条和第 2 条通过执行法案前可以向顾问小组进行咨询协商。欧盟委员会应对该顾问小组的建立和运行制定相关程序。

这些执行法案应符合第二十五节第 2 条的审查程序。

5. 如果某种添加剂或添加剂组合由于成分水平或浓度导致其不符合本节第 1 条的规定而在至少三个成员国被禁止，欧盟委员会有权根据第二十七节采用授权法案，对这些导致特征香味的添加剂或是添加剂组合设定最高限量。

6. 成员国应当禁止含有以下添加剂的烟草制品投放市场：

（a）维生素或其他添加剂，使消费者产生烟草制品有益健康或可以降低健康危害的误解；

（b）与能量和活力有关的咖啡因、牛磺酸或其他添加剂和兴奋剂；

（c）对释放物有着色性能的添加剂；

（d）对于抽吸型烟草制品，能够促进吸入或烟碱摄入的添加剂；以及

（e）在未燃烧状态下具有 CMR 特性的添加剂。

7. 成员国应当禁止在滤嘴、卷烟纸、包装、胶囊等任何组件中

含有香料或是通过技术手段改善烟草制品吸味、吃味或烟气度的烟草制品投放市场。滤嘴、卷烟纸和胶囊中不能含有烟草或烟碱。

8. 各成员国应当确保在 (EC)1907/2006 号法规中制定的条款及条件被酌情应用于烟草制品。

9. 各成员国应当在科学证据的基础上禁止添加剂数量达到将产品毒性、致瘾性或 CMR 特性增加到显著或可测量水平的烟草制品投放市场。

成员国应将依照本条所采取的措施通告欧盟委员会。

10. 欧盟委员会应当在成员国的要求下或是自发地通过执行法案来确定烟草制品是否属于本节第 9 条所指的范围。这些执行法案应符合第二十五节第 2 条的审查程序，并且应当基于最新的科学证据。

11. 如果一种添加剂或一定数量的添加剂表现出烟草制品毒性或致瘾性的放大作用，从而被至少三个成员国依据本节第 9 条禁止，那么欧盟委员会有权根据第二十七节采用授权法案，对这些添加剂设定最高限量，超过此最大限量的产品应当依据本条被其国家禁止。

12. 卷烟和手卷烟之外的烟草制品可以不受本节第 1 条和第 7 条的限制。如果在欧盟委员会报告中发生情况实质性改变，欧盟委员会应当根据第二十七节采用授权法案，撤回对特定产品类别的豁免。

13. 各成员国和欧盟委员会可以向制造商和进口商收取一定比例的费用，用于评估烟草制品是否具有特征香味，是否禁止添加剂或调味剂的使用，以及烟草制品含有的添加剂是否达到能将烟草制品毒性、致瘾性或 CMR 特性增加到显著或可测量水平的数量。

14. 如果某种具有特征香味的烟草制品在欧盟范围内特定产品类别中的销售额在 3% 或以上，则其服从本节规定的时间从 2020 年 5 月 20* 起。

---

* 本指令生效 6 年后。——译注

15. 本节不适用于口用烟草制品。

第二章　标识和包装

第八节　一般规定

1. 烟草制品的每个单位烟包或任何外包装上必须采用官方语言或是产品投放市场所在地成员国的语言标明本章提供的健康警示。

2. 健康警示应当覆盖单位烟包或专为其预留的外包装的整个表面，这些健康警示不能以任何形式被评论、释义或提及。

3. 成员国必须确保单位烟包和任何外包装上的健康警示为永久印刷，不被擦除，且明显可见，包括在烟草制品投放市场时不能被纳税印花、价格标签、防伪标志、包装材料、封皮、盒子或其他部件部分或全部遮挡。在卷烟和袋装手卷烟之外的烟草制品的单位烟包上，健康警示可以用贴纸粘贴，但要保证贴纸无法移动。健康警示应当在打开包装时仍然保持完整，除了采用易拉罐盖子的包装，打开包装时健康警示可能被分开，但仍然必须确保图像的完整性以及文字、图片和戒烟信息的可见性。

4. 健康警示不能被单位烟包上的纳税印花、价格标签、跟踪和追踪标识或防伪标志遮挡或截断。

5. 第九、十、十一和十二节的相关健康警示的尺寸必须根据包装未打开时的表面积来计算。

6. 在警示预留区域内，健康警示必须被围在线宽 1 mm 的黑色边框之内，除了依照第十一节的健康警示。

7. 当依照第九节第 5 条、第十节第 3 条和第十二节第 3 条来修改健康警示时，欧盟委员会必须确认信息是真实的，或者成员国可以在两种警示中做选择，其中之一是真实的。

8. 面向欧盟内消费者的单位烟包和任何外包装的图案均必须遵

循本章的规定。

第九节　抽吸型烟草制品的通用警语和信息

1. 抽吸型烟草制品的每个单位烟包和任何外包装必须带有以下任一通用警语：

“吸烟有害健康——请即刻戒烟”

或

“吸烟有害健康”

成员国可以决定采用哪一条通用警语。

2. 抽吸型烟草制品的每个单位烟包和任何外包装必须带有以下信息：

“烟草烟气中含有 70 多种已知致癌物。”

3. 在卷烟和手卷烟的长方体烟盒上，通用警语必须出现在单位烟包一个侧面的底部，信息则必须出现在另一个侧面的底部。这些健康警示的宽度不能小于 20 mm。

包装采取带有铰链盖的肩盒会导致包装打开时侧面分为两半，通用警语和信息必须完整地出现在分开的两半中表面大的部分，通用警语同时必须出现在上表面的内侧，以便打开时包装时可以看到。

这种包装的侧表面高度应不低于 16 mm。

对于采用袋装销售的手卷烟，通用警语和信息必须出现在表面，以确保这些健康警示的可见性。对于采用圆柱形包装的手卷烟，通用警语必须出现在盖子外表面，信息应出现在盖子内侧。

通用警语和信息必须覆盖所印刷表面的 50%。

4. 本节第 1 条和第 2 条提到的通用警语和信息必须：

（a）采用黑色加粗赫维提卡字体印刷在白色背景上。为适应语言要求，各成员国可以决定字体大小，只要国家法律规定的字体大小确保相关文字占据尽可能多的健康警语

预留表面。以及

（b）位于健康警示预留表面的中心，当印刷在立方体包装及其任何外包装时，应与单位烟包或外包装的侧棱平行。

5. 欧盟委员会有权根据第二十七节采用授权法案，对本节第 2 条的信息修改措辞，以适应科学和市场的发展。

6. 鉴于包装袋形状的差异，欧盟委员会应通过执行法案来决定袋装销售手卷烟上的通用警语和信息的准确位置。

这些执行法案应符合第二十五条第 2 条的审查程序。

第十节　抽吸型烟草制品的组合健康警示

1. 抽吸型烟草制品的每个单位烟包和任何外包装上必须带有组合健康警示。组合健康警示必须：

（a）含有附录一列出的任一文字警语和附录二中的相应彩色图片。

（b）包含戒烟信息，例如电话号码、e-mail 地址或网址，旨在将可行的戒烟方案告知有意戒烟的消费者。

（c）覆盖单位烟包及任何外包装的前后外表面的 65%，圆柱形包装必须展示两条组合健康警示，两条之间等距，各覆盖其半个弧面的 65%。

（d）在单位烟包及任何外包装的两侧展示同样的文字警语和相应的彩色图片。

（e）出现在单位烟包和任何外包装的上边缘，且与出现在包装该表面的其他信息的放置方向相同。出于财政目的仍强制使用纳税印花或国家标识的成员国可以在过渡期间豁免在相应位置印刷组合健康警示的义务，包括以下情况：

（i）出于财政目的的纳税印花或国家标识粘贴在硬盒单位烟包上边缘，将出现在后表面的组合健康警示，可放置在纳税印花或国家识别标志的正下方。

（ii）单位烟包采用软包装时，成员国可以考虑在包装上边缘和组合健康警示的顶端之间预留一个矩形区域，放置出于财政目的的纳税印花或国家识别标志，该矩形区域高度不超过 13 mm。

符合（i）和（ii）条件的豁免期从 2016 年 5 月 20 日 * 开始，为期 3 年。品牌名称和商标不能放置在联合健康警示的上方。

（f）能够根据欧盟委员会依照本节第 3 条指定的格式、布局、设计和比例复制。

（g）卷烟的单位烟包遵循以下尺寸：

（i）高度：不少于 44 mm；

（ii）宽度：不少于 52 mm。

2. 组合健康警示在附录二中分为三组，每一组必须在某一年使用并且每年轮换使用。成员国必须保证每一组组合健康警示在其使用的年度尽可能在每个品牌的烟草制品上显示的次数相同。

3. 欧盟委员会有权根据第二十七节采用以下授权法案：

（a）考虑到科学和市场的发展，修改附录一所列的文字警语；

（b）考虑到科学和市场的发展，建立和修改本节第 1 条（a）中提到的图片库。

4. 鉴于包装外形的差异，欧盟委员会应通过执行法案来定义组合健康警示布局、设计和图形的技术规格。

这些执行法案应符合第二十五节第 2 条的审查程序。

---

* 本指令生效 2 年后。——译注

第十一节　卷烟、手卷烟和水烟之外的抽吸型烟草制品的标识

1. 各成员国可以豁免卷烟、手卷烟和水烟之外的抽吸型烟草制品执行第九节第 2 条要求的信息和第十节要求的组合健康警示的义务。在上述情况下，除第九节第 1 条要求的通用警语外，这些产品的每个单位烟包和任何外包装上必须带有附录一列出的文字警语。第九节第 1 条指定的通用警语必须包含第十节第 1 条（b）中提到的戒烟服务信息。

通用警语必须出现在单位烟包和任何外包装上的最显见表面。

各成员国应确保每条文字警语尽可能在每个品牌的产品上展示的次数相同。文字警语必须出现在单位烟包和任何外包装上的最显见表面的附近。

对于带铰链盖的单位烟包，最显见表面的附近是指包装打开时的可见区域。

2. 本节第 1 条提到的通用警语必须覆盖单位烟包和任何外表面相应表面的 30%。对于使用两种官方语言的成员国，这一比例应提高至 32%，对于使用超过两种官方语言的成员国，这一比例应提高至 35%。

3. 本节第 1 条提到的文字警语必须覆盖单位烟包和任何外包装相应表面的 40%。对于使用两种官方语言的成员国，这一比例应提高至 45%，对于使用超过两种官方语言的成员国，这一比例应提高至 50%。

4. 本节第 1 条提到的健康警示出现在 150 cm^2 以上的包装表面时，这些健康警示必须覆盖 45 cm^2 的面积。对于使用两种官方语言的成员国，该面积应增大至 48 cm^2，对于使用超过两种官方语言的成员国，该面积应增大至 52.5 cm^2。

5. 本节第 1 条提到的健康警示必须符合第九节第 4 条的要求。

健康警示的文本应与健康警示预留区域中的主要文本平行。

健康警示必须被围在线宽 3~4 mm 的黑色边框之内，边框不算在健康警示预留区域里。

6. 如果欧盟委员会报告中关于产品类别划分发生情况实质性改变，则欧盟委员会应根据第二十七节采用授权法案，撤销本节第 1 条提到的对特定产品的豁免权。

第十二节　无烟烟草制品的标识

1. 无烟烟草制品的每个单位烟包和任何外包装上必须带有以下健康警示：

"本烟草制品损害身体健康且具有致瘾性。"

2. 本节第 1 条列出的健康警示必须符合第九节第 4 条的要求。健康警示的文本必须与健康警示预留区域中的主要文本平行。

另外，健康警示应：

（a）出现在单位烟包和任何外包装的最大的两个表面；

（b）覆盖单位烟包和任何外包装表面的 30%。对于使用两种官方语言的成员国，这一比例应提高至 32%，对于使用超过两种官方语言的成员国，这一比例应提高至 35%。

3. 欧盟委员会有权根据第二十七节采用授权法案，修改本节第 1 条指定的健康警示，以适应科学的发展。

第十三节　产品描述

1. 烟草制品单位烟包和任何外包装上的标识及烟草制品本身不能包含以下任何要素和特征：

（a）通过建立烟草制品特征、健康影响、风险性或释放物的错误印象来推广产品或鼓励消费；标识中不能包含关于烟草制品烟碱、焦油或一氧化碳含量的任何信息。

（b）暗示一种特定烟草制品的危害低于其他烟草制品，或者以降低烟气有害物质的影响为目的，或者含有激发、振奋、治愈、修护、天然、有机等特征，或者对健康和生活品质有益处。

（c）提及口味、气味、任何香味或其他添加剂，或者注明不含这些添加剂。

（d）类似于食品或美容产品。

（e）暗示特定烟草制品有较好的生物降解能力或其他环境益处。

2. 烟草制品单位烟包和任何外包装都不能提供代金券、折扣、免费派发、买一送一或其他类似的促销活动。

3. 本节第 1 条和第 2 条所禁止的要素和特征包含但不限于文字、符号、名称、商标、图形或其他标志。

第十四节　单位烟包的外观和内容

1. 卷烟的单位烟包应为长方体。手卷烟的单位烟包应为长方体或圆柱体，或者以袋装的形式。卷烟的单位烟包内应至少含有 20 支卷烟。手卷烟的单位烟包内含有的烟草应不少于 30 g。

2. 卷烟的单位烟包可由硬纸盒或软材料制成，不能含有第一次打开后可以重新关闭或重新密封的开口，除了易拉罐包装和带有铰链盖的肩盒。对于易拉罐盖和铰链盖，盖子应在单位烟包的背面装上铰链。

第十五节　可追溯性

1. 各成员国应该确保烟草制品的所有单位烟包上有一个特有标识符。为了确保特有标识符的完整性，它应被不可去除地打印或粘贴，不能擦除且不被遮挡或被纳税印花、价格标签或单位烟包的开口等任何形式遮挡或截断。在欧盟以外制造的烟草制品，本节规定的义务仅适用于那些将要或已经投放到欧盟市场的产品。

2. 特有标识符应考虑以下待定项目：

（a）生产日期和地点；

（b）制造设施；

（c）用于制造烟草制品的机器；

（d）生产轮班或制造时间；

（e）产品描述；

（f）目标零售市场；

（g）预定的装运路线；

（h）在适用的情况下，引入欧盟的进口商；

（i）从生产到第一个零售商店的实际装运路线，包括所有的仓库和装运日期、目的地、出货时间和收货人；

（j）从生产到第一个零售商店的所有购买者的身份；以及

（k）从生产到第一个零售商店的所有购买者的发票、订单号和付款记录。

3. 本节第 2 条（a），（b），（c），（d），（e），（f），（g）提到的信息，以及适用情况下（h）的信息应构成特有标识符的部分内容。

4. 各成员国应确保本节第 2 条第（i），（j）和（k）提到的信息可通过与特有标识符链接实现电子访问。

5. 各成员国应确保所有参与烟草制品贸易的经济运营商（从制造商到第一个零售商店之前的最后一个经济运营商）记录进入他们账户的所有单位烟包，以及从其账户中转和最终出口的单位烟包。该义务可与条、箱或件等大批量包装的标记和记录相兼容，只要保持对所有单位烟包的跟踪和追踪。

6. 各成员国应确保烟草制品供应链上的所有自然人和法人对所有相关交易保留完整和精确的记录。

7. 各成员国应确保烟草制品制造商提供的所有涉及烟草制品贸易的经济运营商，从制造商到第一个零售商店之前的最后一个经济运营商，包括进口商、仓储和运输公司，应配备能记录烟草制品购买、销售、储存、运输或其他处理数据的必要的设备。该设备应该可以读取记录的电子数据并发送到本节第 8 条提到的数据存储设备。

8. 各成员国应确保烟草制品的制造商和进口商与独立的第三方签订数据存储合约，以托管所有相关数据的数据存储设备。数据存储设备必须位于欧盟地区。第三方的适用性，尤其是其独立性和技术能力，以及数据存储合约，必须获得欧盟委员会的批准。

第三方的活动需要由外部审计师进行监测，该外部审计师由烟草制造商推荐和付酬，并且经过欧盟委员会的批准。外部审计师须每年向主管部门以及欧盟委员会提交报告，特别是评估任何关于访问的违规行为。

各成员国应确保欧盟委员会、各成员国的主管部门和外部审计师能完全访问数据存储设备。在适当合理的情况下，欧盟委员会和各成员国可以允许制造商或进口商访问存储数据，前提是依照欧盟和国家相关法律对商业敏感信息进行充分的保护。

9. 记录的数据不得由参与烟草制品贸易的经济运营商修改或删除。

10. 各成员国应确保个人数据的处理与 95/46/EC 指令中设定的规则和保护措施相一致。

11. 欧盟委员会应通过执行法案：

（a）为本节提供的跟踪和追踪系统的创立和实施确定技术标准，包括有特有标识符的标志、记录、传输、处理和数据存储以及访问存储的数据；

（b）为确保使用的特有识别符系统及相关功能在整个欧盟内具有充分兼容性而制定技术标准。

这些执行法案应符合第二十五节第 2 条的审查程序。

12. 欧盟委员会应根据第二十七节采用授权法案，定义本节第 8 条中涉及的数据存储合约的关键要素，如持久性、可再生性、专业性要求或机密性，包括对那些合同的定期监测和评价。

13. 从 2019 年 5 月 20 日 * 开始，本节第 1~10 条应适用于卷烟和手卷烟，从 2024 年 5 月 20 日 ** 开始，本节第 1~10 条应适用于卷烟和手卷烟之外的烟草制品。

第十六节　防伪标志

1. 除了第十五节提到的特有标识符以外，各成员国应要求所有投放市场的烟草制品的单位烟包带有防篡改的防伪标志，包括可见和不可见的要素。防伪标志应被不可移动地打印或粘贴，不可除去，且不被纳税印花、价格标签或立法要求的其他要素等遮挡或截断。

成员国要求的出于财政目的纳税印花或国家标识也被允许用于防伪标志，只要纳税印花或国家标识满足本节要求的技术标准和功能。

2. 欧盟委员会应通过执行法案，为防伪标志及其可能的轮换方式制定技术标准，并根据科学、市场和技术的发展对技术标准进行修订。

这些执行法案应符合第二十五节第 2 条的审查程序。

3. 从 2019 年 5 月 20 日 * 开始，本节第 1 条应适用于卷烟和手卷烟，从 2024 年 5 月 20 日 ** 开始，本节第 1 条应适用于卷烟和手卷烟之外的烟草制品。

---

* 本指令生效 5 年后。——译注

** 本指令生效 10 年后。——译注

第三章　口用烟制品草，烟草制品的跨境远程销售及新型烟草制品

第十七节　口用烟草制品

在不损害奥地利、芬兰和瑞典加入的法令的第 151 条要求的前提下，各成员国应禁止口用烟草制品投放市场。

第十八节　烟草制品的跨境远程销售

1. 各成员国可以禁止烟草制品跨境远程销售给消费者。各成员国应互相协作以阻止此类销售。从事烟草制品跨境远程销售的零售商不应向已禁止烟草制品跨境远程销售成员国的消费者提供这类产品。不禁止此类销售的成员国应要求有意涉足跨境远程销售给欧盟消费者的零售商在其所在成员国，以及实际或潜在消费者所在成员国的主管部门登记。位于欧盟之外的零售商应被要求在实际或潜在消费者所在成员国的主管部门登记。在注册时，所有有意涉足跨境远程销售的零售商应至少向主管部门提交如下信息：

（a）品牌名称或公司名称以及烟草制品供应场所的永久地址；

（b）通过信息社会服务向消费者提供跨境远程销售的烟草制品的开始日期，参照 98/34/EC 指令的第 1（2）条的定义；

（c）用于此目的的一个或多个网站的网址以及能识别该网址的所有必要相关信息。

2. 各成员国的主管部门应确保消费者可以获取所有注册的零售商清单。当使该清单可被获取时，各成员国应确保遵守 95/46/EC 指令设定的规则和保障措施。当登记获得相关主管部门的批准后，零售商才可以开始通过跨境远程销售将烟草制品投放市场。

3. 如果为了确保遵守法规和促进法规实施有必要进行验证的话，烟草制品跨境远程销售的目的地成员国可以要求供货的零售商推荐一位自然人，负责在烟草制品到达消费者之前的验证工作，以验证

该烟草制品遵守销售目的地成员国依据本指令所出台的国家规定。

4. 跨境远程销售的零售商应实行年龄认证系统，以在销售时验证购买者是否符合销售目的地成员国国家法律规定的最低年龄要求。零售商或根据本节第 3 条推荐的自然人应向成员国主管部门提供关于年龄认证系统的细节和功能描述。

5. 零售商应按照 95/46/EC 指令要求仅处理消费者的个人数据，并且这些数据不得向烟草制品制造商或同类公司的子公司或其他第三方披露。除了实际购买之外，个人数据不得用于其他目的的使用或转让。当零售商是烟草制品制造商的一部分时，本条同样适用。

第十九节　关于新型烟草制品的通告

1. 各成员国应要求新型烟草制品的制造商和进口商向他们想要将此类产品投放市场的相关成员国的主管部门提交通告。通告应以电子形式在产品投放市场之前 6 个月提交。它应该附带一个对该新型烟草制品的详细描述和使用说明，以及遵照第五节要求的成分和释放物信息。制造商和进口商向主管部门提交的关于新烟草制品的通告还应含有以下信息：

（a）可获得的关于新型烟草制品毒性、致瘾性和吸引力的科学研究，尤其是涉及其成分和释放物的科学研究；

（b）可获得的关于不同消费群体（包括青少年和当前吸烟者）偏好的研究及得出的总结报告和市场调查；

（c）其他可获得的相关信息，包括产品的风险 / 益处分析，其对戒烟的预期效果，其对消费者开始使用烟草制品的预期效果以及预期的消费者认知。

2. 各成员国应要求新型烟草制品的制造商和进口商向主管部门提交和更新任何新的研究、调查以及本节第 1 条（a）~（c）中所提及的其他信息。成员国可要求新型烟草制品的制造商或进口商进行额外的测试或提交额外的信息。成员国应根据本节要求收集所有相关信息，并提供给欧盟委员会使用。

3. 成员国可以引入新型烟草制品的授权系统。成员国可以向制造商和进口商收取一定比例的费用用于该授权工作。

4. 投放市场的新烟草制品应遵守本指令的要求。本指令的哪些规定适用于新型烟草制品，取决于该产品属于无烟烟草制品的定义范畴还是抽吸型烟草制品的定义范畴。

第三部分　电子烟和抽吸型草本制品

第二十节　电　子　烟

1. 各成员国应确保只有符合本指令和其他相关欧盟立法的电子烟和贮液容器才能投放市场。

本指令不适用于满足 2001/83/EC 指令中的授权要求或 93/42/EC 指令要求的电子烟和贮液容器。

2. 电子烟和贮液容器的制造商和进口商应向他们想要将此类产品投放市场的相关成员国的主管部门提交通告。通告应以电子形式在产品投放市场之前 6 个月提交。对于在 2016 年 5 月 20 日 * 前已投放市场的电子烟和贮液容器，通告应在 2016 年 5 月 20 日后的 6 个月内提交。每当产品发生实质性改变时都应提交新的通告。

通告应根据产品是否属于电子烟或贮液容器而含有以下信息：

（a）制造商的名称和详细联系方式，欧盟范围内的负有责任的法人或自然人，以及在适用时将产品引入欧盟的进口商；

（b）按品牌名称和产品类型划分的产品所有成分清单，以及使用该产品后产生的释放物，包括其含量；

* 本指令生效 2 年后。——译注

（c）关于产品成分和释放物的毒理学数据，包括加热后的毒理学数据，尤其当吸入后对消费者的健康影响以及成瘾影响；

（d）烟碱剂量和消费者在正常使用或者合理预测条件下的烟碱摄入量相关信息；

（e）产品组成的描述，包括在适用的情况下电子烟或贮液容器的开启和填充烟液的原理；

（f）生产过程描述，包括是否涉及批量生产，以及是否有确保生产过程与本节要求相一致的声明；

（g）投放市场的产品在正常使用或合理预测条件下，制造商和进口商对产品质量和安全负全责的声明。

当成员国认为提交的信息不完整时，他们有权要求完善相关信息。

成员国可以向制造商和进口商收取一定比例的费用，用于接收、储存、处理和分析所提交的信息。

3. 成员国应确保：

（a）含有烟碱的烟液仅在专用的贮液容器内投放市场，且贮液容器的体积不超过 10 mL，在一次性电子烟或一次性烟弹中，烟弹或烟液池的体积不超过 2 mL。

（b）含有烟碱的烟液中，烟碱含量不得超过 20 mg/mL。

（c）含有烟碱的烟液中，不得含有第七节第 6 条所列出的添加剂。

（d）仅高纯度的成分才能用于制造含有烟碱的烟液。本节第 2 条（b）中涉及的成分之外的物质仅能以痕量水平存在于含有烟碱的烟液中，如果这种痕量残留在制造过程中是技术上不可避免的。

（e）除烟碱之外，只有那些在加热或不加热情况下都不会对人体健康构成风险的成分才能使用到含有烟碱的烟液中。

（f）在正常使用条件下，电子烟的烟碱传输量应维持在稳定水平。

（g）电子烟和贮液容器应含有儿童防护和改装防护功能，应防止破碎或泄露，且含有确保在烟液填充过程中不漏液的设计。

4. 成员国必须确保：

（a）电子烟和贮液容器的单位烟包含有涉及以下信息的说明书：

（i）产品使用和储存的说明书，包括该产品不适合青少年和非吸烟者使用的建议；

（ii）使用禁忌；

（iii）对特定高危人群的警告；

（iv）可能产生的副作用；

（v）致瘾性和毒性；以及

（vi）制造商或进口商的联系方式以及在欧盟范围内法人或自然联系人的联系方式；

（b）电子烟和贮液容器的单位烟包和任何外包装：

（i）包括按照含量降序排列的所有产品成分的列表，产品烟碱含量及单位剂量所产生的烟碱传输量的说明，批号和该产品放置在儿童不可触及区的建议；

（ii）与（i）点不冲突，除第十三节第1条（a）和（c）中关于烟碱含量和调味剂的相关信息外，不包括第十三节提及的要素和特征；以及

（iii）具有以下任一健康警示：

“本产品含有强致瘾性物质——烟碱，不建议非吸烟者使用。”

或

“本产品含有烟碱，一种强致瘾性物质。”

成员国应确定在上述健康警示中选择哪条使用。

（c）健康警示与第十二节第 2 条的要求相一致。

5. 各成员国应确保：

（a）禁止以直接或间接促进电子烟和贮液容器销售为目的，通过信息社会服务、图书或其他印刷出版物进行商业传播，除了专门针对电子烟或贮液容器贸易的专业人士的出版物，以及由第三方国家印刷和出版的出版物，且这些出版物并非主要面向欧盟市场；

（b）禁止以直接或间接促进电子烟和贮液容器销售为目的，通过电台进行商业传播；

（c）禁止以直接或间接促进电子烟和贮液容器销售为目的的任何形式的公共或私人赞助的广播节目；

（d）禁止以直接或间接促进电子烟和贮液容器销售为目的，发生在多个成员国或有跨境影响的，任何形式的公共或私人赞助的事件、活动或个人；

（e）欧洲议会和欧盟理事会 2010/13/EU 指令[(1)]中适用的视听商业传播，被禁止用于电子烟和贮液容器。

6. 本指令第十八节应适用于电子烟和贮液容器的跨境远程销售。

7. 各成员国应要求电子烟和贮液容器的制造商和进口商每年向主管部门提交以下信息：

（i）按品牌名称和产品类型划分的销售量的综合数据；

（ii）不同消费群体（包括青少年、不吸烟者和当前吸烟者的

---

(1) 2010 年 3 月 10 日欧洲议会和欧盟理事会关于协调各成员国有关视听媒体服务规定的法律、法规和管理规定（“视听媒体服务指令”）的 2010/13/EU 指令 (OJ L 95, 15.4.2010, p. 1)。

主要类型）的偏好信息；

（iii）产品的销售模式；以及

（iv）关于以上方面开展的任何社会调查的总结，包括其中的英文译本。

各成员国应监视关于电子烟和贮液容器的市场发展，包括证明它们的使用是导致青少年和非吸烟者烟碱成瘾并最终发展为传统烟草消费的途径的任何证据。

8. 各成员国应确保按照本节第 2 条提交的信息通过网站向公众发布。当信息公开时，各成员国应充分考虑保护商业机密。

若有需求，各成员国应将依照本节收集的所有信息提供给欧盟委员会或其他成员国。各成员国和欧盟委员会应确保商业机密和其他机密信息按照保密模式处理。

9. 各成员国应要求电子烟和贮液容器的制造商、进口商和分销商建立和维持一个用于收集此类产品的所有可能的不良健康影响的信息收集系统。

若有任何经济运营商认为或有理由相信他们拥有的和即将投放市场的或已经投放市场的电子烟和贮液容器不安全或质量不好或与本指令规定不相符，则经济运营商应根据具体情况立即采取必要的纠正措施使相关产品符合本指令要求，或者撤销或召回相关产品。在这种情况下，经济运营商同时应立即告知产品所在或拟投放市场的成员国的监管部门相关细节，尤其是对人体健康和安全的风险以及所采取的所有纠正措施以及这些纠正措施所带来的结果。

成员国也可从经济运营商那里要求额外的信息，如电子烟或贮液容器的安全性和质量方面或任何不利影响。

10. 欧盟委员会应在 2016 年 5 月 20 日 * 前及之后的适当时间向

---

* 本指令生效 2 年后。——译注

欧洲议会和欧盟理事会提交关于可填充型电子烟对公共健康潜在风险的报告。

11. 在电子烟和贮液容器符合本节要求的情况下，当主管部门确定或有充分理由相信特定电子烟或贮液容器，或者一类电子烟或贮液容器，可以对人体健康产生严重的风险时，可以采取适当的临时措施。应立即告知欧盟委员会和其他成员国的主管部门所采取的措施，并应交流支持数据。欧盟委员会在收到信息后，应尽快判定所采取的临时措施是否得当。欧盟委员会还应将该结论告知相关成员国，使其能采取适当的后续措施。

在本条第 1 段的应用中，若特定电子烟或贮液容器，或者一类电子烟或贮液容器在至少三个成员国以合理理由禁止投放市场时，欧盟委员会有权根据第二十七节采用授权法案，将此禁令扩大至所有的成员国，如果这种扩大是合理和适当的。

12. 欧盟委员会有权根据第二十七节采用授权法案，修改本节第 4 条（b）中的健康警示。当修改健康警示时，欧盟委员会应确保其真实性。

13. 欧盟委员会应通过执行法案，为第 2 条要求的通告制定统一格式并为第 3 条（g）要求的烟液填充原理设定技术标准。

这些执行法案应符合第二十五节第 2 条的审查程序。

第二十一节　抽吸型草本制品

1. 抽吸型草本制品的每个单位烟包和任何外包装都应带有以下健康警示：

“抽吸本产品有害健康。”

2. 健康警示应印刷在单位烟包的前后外表面和任何外包装上。

3. 健康警示应与第九节第 4 条的要求相一致。它应覆盖单位烟包和任何外包装相应表面的 30%。对于使用两种官方语言的成员国，该比例应增至32%，对于使用超过两种官方语言的成员国，该比例应增至35%。

4. 抽吸型草本制品的单位烟包和任何外包装应不包括第十三节第1条（a），（b）和（d）中设立的要素和特征，且不得声明该产品不含添加剂或调味剂。

第二十二节　抽吸型草本制品的成分报告

1. 各成员国应要求抽吸型草本制品的制造商和进口商向主管部门提交按品牌名称和产品类型划分的此类产品在生产过程中使用的所有成分清单及其含量。当产品的组成发生改变且影响到按本节要求提交的信息时，制造商或进口商也应告知相关成员国主管部门。本节要求的信息应在新的或改良的抽吸型草本制品投放市场之前提交。

2. 各成员国应确保按照第1条提交的信息通过网站向公众发布。当信息公开时，各成员国应充分考虑保护商业机密。经济运营商应明确指出哪些信息他们认为属于商业机密。

第四部分　最 后 条 款

第二十三节　协作与执行

1. 各成员国应确保烟草及其相关产品的制造商和进口商按照本指令要求在限定时间内向欧盟委员会和成员国主管部门提供完整和正确的信息。如果制造商设立在欧盟区域，则提供所要求信息的义务应主要在于制造商。如果制造商设立在欧盟之外，而进口商设立在欧盟区域，则提供所要求信息的义务主要在于进口商。如果制造商和进口商都设立在欧盟之外，则提供所要求信息的义务由制造商和进口商共同承担。

2. 成员国应确保不满足本指令要求（包括其中涉及的执行法案和授权法案）的烟草及其相关产品不能投放市场。如果不能满足本指令要求的报告义务，则成员国应确保该烟草及其相关产品不得投放市场。

3. 各成员国应制定对违反依据本指令采取的国家法律的处罚规则，并且采取所有必要措施以确保处罚的强制执行。提供的处罚应该是有效的、适当的和有劝阻性的。当故意违规发生时，可施以经济行政处罚，或许可抵消通过违规获得的经济利益。

4. 各成员国的主管部门应彼此间相互合作，并应与欧盟委员会合作，以确保本指令的准确应用和如期执行，还应彼此间传递所有必要信息从而使本指令按统一的方式执行。

第二十四节　自 由 流 通

1. 考虑到本指令管制的相关方面及本节第 2 条和第 3 条的内容，成员国可以不禁止或限制符合本指令要求的烟草及其相关产品投放市场。

2. 本指令应不影响成员国针对投入其市场的所有产品，在烟草制品包装的标准化方面，维持或采取进一步要求的权利，只要该要求合理地以公众健康为理由，考虑到通过本指令获得的高水平的人体健康保护。这些措施应适当，且不能构成恶意歧视手段或成员国之间的变相贸易壁垒。这些措施应与维持或采取的理由一起通告欧盟委员会。

3. 成员国也可以基于本国具体情况禁止特定类别的烟草及其相关产品，只要该规定从出于保护公众健康需要的角度是合理的，同时考虑到通过本指令获得的高水平的人体健康保护。此类国家法规应与引进该法规的理由一起通告欧盟委员会。欧盟委员会应在接收到本条所指的通告之日起的6个月内，进行如下审核：考虑到通过本指令获得的高水平的人体健康保护，来评价该国家法规对于上述目标而言是否合理、必要和合适，以及该国家法规是否在成员国之间构成恶意歧视和变相贸易壁垒，进而决定批准或驳回该国家法规。如果欧盟委员会在6个月内未作出决定，则该国家法规被视为获得批准。

第二十五节　委员会程序

1. 欧盟委员会须有一个委员会协助。这个委员会应符合（EU）182/2011号法规。

2. 本条的规定参考（EU）182/2011号法规第5条。

3. 委员会的意见通过书面程序获得，在传达该意见的期限内，当委员会主席决定或大多数委员会成员请求时，该程序将以无结果状态被终止。

4. 若委员会没有结论意见，则欧盟委员会不得采用执行法案草案，而应适用（EU）182/2011号法规第5（4）条第3段内容。

第二十六节　主管部门

各成员国应在2016年5月20日*之后的3个月内，指定主管部门负责本指令规定义务的实施和执行。各成员国应立即告知欧盟委员会指定主管部门的身份。欧盟委员会应在欧盟官方杂志上公布该信息。

第二十七节　授权运用

1. 在服从本指令要求的情况下，授予欧盟委员会采用授权法案的权利。

---

*　本指令生效2年后。——译注

2. 自 2014 年 5 月 19 日 *起的 5 年内，涉及第三节第 2 条和第 4 条，第四节第 3 条和第 5 条，第七节第 5 条、第 11 条和第 12 条，第九节第 5 条，第十节第 3 条，第十一节第 6 条，第十二节第 3 条，第十五节第 12 条，第二十节第 11 条和第 12 条的内容，授予欧盟委员会采用授权法案的权利。在 5 年期结束前的 9 个月之前，欧盟委员会应关于授权权利起草一个报告。授权权利将默认延长一段时间，除非欧洲议会或欧盟理事会在该 5 年期结束前的 3 个月之前反对该延期。

3. 涉及第三节第 2 条和第 4 条，第四节第 3 条和第 5 条，第七节第 5 条、第 11 条和第 12 条，第九节第 5 条，第十节第 3 条，第十一节第 6 条，第十二节第 3 条，第十五节第 12 条，第二十节第 11 条和第 12 条的授权权利可随时被欧洲议会或欧盟理事会取消。取消的决定应明确结束哪项授权权利。在欧盟官方杂志上公布后或在稍后指定的日期中，该决定将生效。它不影响任何已执行授权法案的有效性。

4. 一旦采用授权法案，欧盟委员会应同时通告欧洲议会和欧盟理事会。

5. 依据第三节第 2 条和第 4 条，第四节第 3 条和第 5 条，第七节第 5 条、第 11 条和第 12 条，第九节第 5 条，第十节第 3 条，第十一节第 6 条，第十二节第 3 条，第十五节第 12 条，第二十节第 11 条和第 12 条采取授权法案的生效条件是：在将该法案通告欧洲议会和欧盟理事会的 2 个月内，欧洲议会或欧盟理事会没有表示反对；或在 2 个月期满之前，欧洲议会和欧盟理事会同时告知欧盟委员会他们没有反对意见。欧洲议会或欧盟理事会可以倡议将这个期限延长 2 个月。

---

* 本指令生效日期。——译注

第二十八节　报　　告

1. 自 2016 年 5 月 20 日 * 开始的 5 年内，在任何必要的时候，欧盟委员会应向欧洲议会、欧盟理事会、欧洲经济和社会委员会以及各地区委员会提交关于本指令执行情况的报告。

当起草该报告时，欧盟委员会应由科学家和技术专家协助，以使得所有必要的信息得到合理处置。

2. 在报告中，欧盟委员会应尤其标明本指令中哪些元素应被审查或随科学和技术的发展进行改进，包括关于烟草及其相关产品的国际通用规则和标准的发展。欧盟委员会应特别注意：

（a）考虑到国家、国际、法律、经济和科学的发展，关于外包装设计的经验不受本指令的管制；

（b）关于新型烟草制品的市场发展，尤其是按第十九节要求接收的通告；

（c）发生情况实质性改变的市场发展；

（d）欧盟对烟草制品所用成分进行管控的可行性、效益和可能的影响，包括在欧盟层面设立的在烟草制品中使用、存在或添加的成分清单，尤其考虑到第五节和第六节对信息收集的要求；

（e）关于直径小于 7.5 mm 的卷烟的市场发展，对其危害性的消费者认知和此类卷烟的误导性；

（f）含有根据第五节和第六节要求收集的烟草制品成分和释放物信息的欧盟数据库的可行性、效益和可能的影响；

（g）关于电子烟和贮液容器的市场发展，根据第二十节要求收集到的信息，包括青少年和非吸烟者开始消费此类产品的信息，此类产品的戒烟功效以及各成员国关于调味

---

* 本指令生效 2 年后。——译注

剂所采取的措施；

（h）关于水烟的市场发展和消费者偏好，尤其关注其香味。

各成员国应协助欧盟委员会提供所有可用的信息，以开展评估和准备报告。

3. 如果欧盟委员会认为，为实现境内市场的平稳运作，为适应烟草及其相关产品领域的发展，并考虑到基于科学事实的新发展和烟草及其相关产品国际通用标准的发展，有必要对本指令进行修订，则该报告应作为本指令修订提案的附件。

第二十九节　转　　换

1. 各成员国应按本指令要求在 2016 年 5 月 20 日 * 之前实施必要的法律、法规和管理规定，并应立即将上述规定的文本告知欧盟委员会。

在不妨碍第七节第 14 条，第十节第 1 条（e），第十五节第 13 条和第十六节第 3 条的情况下，各成员国应从 2016 年 5 月 20 日 * 开始实施相关措施。

2. 当成员国制定相关规定时，这些规定应包括参考的本指令的相关内容，或在其官方出版物中附上相关参考内容。还应有一个声明：现有法律、法规和管理规定参考了被本指令废除的指令中的哪些内容，以及涉及本指令中的哪些相关参考内容。成员国应决定如何准备这些参考内容以及怎样阐述该声明。

3. 成员国应向欧盟委员会提交在本指令覆盖领域内所采取的国家法律主要规定的文本。

---

*　本指令生效 2 年后。——译注

第三十节　过渡性条款

成员国可允许不符合本指令要求的以下产品投入市场直到2017年5月20日*为止：

（1）在2016年5月20日**之前，其生产或自由流通、标识均符合2001/37/EC指令要求的烟草制品；

（2）在2016年11月20日***之前已生产或自由流通的电子烟或贮液容器；

（3）在2016年5月20日**之前已生产或自由流通的抽吸型草本制品。

第三十一节　废　　止

在不妨碍各成员国将本指令转换至国家法律的相关时限义务的情况下，自2016年5月20日**起，2001/37/EC指令被废止。

被废止指令的参考内容应被解释为本指令的参考内容，可参照本指令附录三的对照表。

第三十二节　生　　效

本指令自在欧盟官方杂志上发表后的第20天起生效。

第三十三节　适 用 对 象

本指令的适用对象为欧盟各成员国。

---

*　本指令生效3年后。——译注

**　本指令生效2年后。——译注

***　本指令生效30个月后。——译注

2014 年 4 月 3 日完成于布鲁塞尔。

欧洲议会	欧盟理事会
主席	理事长
M. Schulz	D. Kourkoulas

附录一　文字警语列表

（参照第十节和第十一节第1条）

（1）吸烟使10人中的9人罹患肺癌

（2）吸烟导致口腔癌和咽喉癌

（3）吸烟损害你的肺

（4）吸烟引发心脏病

（5）吸烟导致中风和残疾

（6）吸烟使你的动脉堵塞

（7）吸烟增加失明的危险

（8）吸烟损害你的牙齿和牙龈

（9）吸烟会杀死你未出世的孩子

（10）吸烟危害你的孩子、家人和朋友

（11）吸烟者的孩子更容易开始吸烟

（12）戒烟——让你身边的人活下去

（13）吸烟导致生育率降低

（14）吸烟增加性功能障碍的风险

附录二　图　片　库

（参照第十节第 1 条）

[由欧盟委员会根据第十节第 3 条（b）设置]

附录三　对　照　表

2001/37/EC 指令	本指令
第一节	第一节
第二节	第二节
第三节第 1 条	第三节第 1 条
第三节第 2 条和第 3 条	—
第四节第 1 条	第四节第 1 条
第四节第 2 条	第四节第 2 条
第四节第 3~5 条	—
第五节第 1 条	—
第五节第 2 条（a）	第九节第 1 条
第五节第 2 条（b）	第十节第 1 条（a）和第 2 条，第十一节第 1 条
第五节第 3 条	第十节第 1 条
第五节第 4 条	第十二节
第五节第 5 条第 1 段	第九节第 3 条第 5 段 第十一节第 2 条和第 3 条 第十二节第 2 条（b）
第五节第 5 条第 2 段	第十一节第 4 条
第五节第 6 条（a）	第九节第 4 条（a）
第五节第 6 条（b）	—
第五节第 6 条（c）	第九节第 4 条（b）
第五节第 6 条（d）	第八节第 6 条和第十一节第 5 条第 2 段
第五节第 6 条（e）	第八节第 1 条
第五节第 7 条	第八节第 3 条和第 4 条
第五节第 8 条	—
第五节第 9 条第 1 段	第十五节第 1 条和第 2 条
第五节第 9 条第 2 段	第十五节第 11 条
第六节第 1 条第 1 段	第五节第 1 条第 1 段
第六节第 1 条第 2 段	第五节第 2 条和第 3 条
第六节第 1 条第 3 段	—
第六节第 2 条	第五节第 4 条
第六节第 3 条和第 4 条	—
第七节	第十三节第 1 条（b）
第八节	第十七条
第九节第 1 条	第四节第 3 条
第九节第 2 条	第十节第 2 条和第 3 条（a）

2001/37/EC 指令	本指令
第九节第 3 条	第十六节第 2 条
第十节第 1 条	第二十五节第 1 条
第十节第 2 条和第 3 条	第二十五节第 2 条
第十一节第 1 段和第 2 段	第二十八节第 1 条第 1 段和第 2 段
第十一节第 3 段	第二十八节第 2 条第 1 段
第十一节第 4 段	第二十八节第 3 条
第十二节	—
第十三节第 1 条	第二十四节第 1 条
第十三节第 2 条	第二十四节第 2 条
第十三节第 3 条	—
第十四节第 1 条第 1 段	第二十九节第 1 条第 1 段
第十四节第 1 条第 2 段	第二十九节第 2 条
第十四节第 2 条和第 3 条	第三十节（a）
第十四节第 4 条	第二十九节第 3 条
第十五节	第三十一节
第十六节	第三十二节
第十七节	第三十三节
附录一 附加健康警示列表	附录一 文字警语列表
附录二 对废止指令的转换和执行时限	—
附录三 对照表	附录三 对照表

I

(Legislative acts)

DIRECTIVES

DIRECTIVE 2014/40/EU OF THE EUROPEAN PARLIAMENT AND OF THE COUNCIL

of 3 April 2014

on the approximation of the laws, regulations and administrative provisions of the Member States concerning the manufacture, presentation and sale of tobacco and related products and repealing Directive 2001/37/EC

(Text with EEA relevance)

THE EUROPEAN PARLIAMENT AND THE COUNCIL OF THE EUROPEAN UNION,

Having regard to the Treaty on the Functioning of the European Union, and in particular Articles 53(1), 62 and 114 thereof,

Having regard to the proposal from the European Commission,

After transmission of the draft legislative act to the national parliaments,

Having regard to the opinion of the European Economic and Social Committee(1),

Having regard to the opinion of the Committee of the Regions(2),

(1) OJ C 327, 12.11.2013, p. 65.

(2) OJ C 280, 27.9.2013, p. 57.

Acting in accordance with the ordinary legislative procedurex[1],

Whereas:

(1) Directive 2001/37/EC of the European Parliament and of the Council[2] lays down rules at Union level concerning tobacco products. In order to reflect scientific, market and international developments, substantial changes to that Directive would be needed and it should therefore be repealed and replaced by a new Directive.

(2) In its reports of 2005 and 2007 on the application of Directive 2001/37/EC the Commission identified areas in which further action was considered useful for the smooth functioning of the internal market. In 2008 and 2010 the Scientific Committee on Emerging and Newly Identified Health Risks (SCENIHR) provided scientific advice to the Commission on smokeless tobacco products and tobacco additives. In 2010 a broad stakeholder consultation took place, which was followed by targeted stakeholder consultations and accompanied by studies by external consultants. Member States were consulted throughout the process. The European Parliament and the Council repeatedly called on the Commission to review and update Directive 2001/37/EC.

(3) In certain areas covered by Directive 2001/37/EC, Member States are legally or in practice prevented from effectively adapting their legislation to new developments. This is in particular relevant for the labelling rules, where Member States have not been permitted to increase the size of the health warnings, change their location on an individual packet ('unit packet') or replace misleading warnings on the tar, nicotine and carbon monoxide (TNCO) emission levels.

(4) In other areas there are still substantial differences between the Member States' laws, regulations and administrative provisions on the manufacture, presentation and sale of tobacco and related products which present obstacles to the smooth functioning of the internal market. In the light of scientific, market and international developments these discrepancies are expected to increase. This also

---

(1) Position of the European Parliament of 26 February 2014 (not yet published in the Official Journal) and decision of the Council of 14 March 2014.

(2) Directive 2001/37/EC of the European Parliament and of the Council of 5 June 2001 on the approximation of the laws, regulations and administrative provisions of the Member States concerning the manufacture, presentation and sale of tobacco products (OJ L 194, 18.7.2001, p. 26).

applies to electronic cigarettes and refill containers for electronic cigarettes ('refill containers'), herbal products for smoking, ingredients and emissions from tobacco products, certain aspects of labelling and packaging and to cross-border distance sales of tobacco products.

(5) Those obstacles should be eliminated and, to this end, the rules on the manufacture, presentation and sale of tobacco and related products should be further approximated.

(6) The size of the internal market in tobacco and related products, the increasing tendency of manufacturers of tobacco products to concentrate production for the entire Union in only a small number of production plants within the Union and the resulting significant cross-border trade of tobacco and related products calls for stronger legislative action at Union rather than national level to achieve the smooth functioning of the internal market.

(7) Legislative action at Union level is also necessary in order to implement the WHO Framework Convention on Tobacco Control ('FCTC') of May 2003, the provisions of which are binding on the Union and its Member States. The FCTC provisions on the regulation of the contents of tobacco products, the regulation of tobacco product disclosures, the packaging and labelling of tobacco products, advertising and illicit trade in tobacco products are particularly relevant. The Parties to the FCTC, including the Union and its Member States, adopted a set of guidelines for the implementation of FCTC provisions by consensus during various Conferences.

(8) In accordance with Article 114(3) of the Treaty of the Functioning of the European Union (TFEU), a high level of health protection should be taken as a base for legislative proposals and, in particular, any new developments based on scientific facts should be taken into account. Tobacco products are not ordinary commodities and in view of the particularly harmful effects of tobacco on human health, health protection should be given high importance, in particular, to reduce smoking prevalence among young people.

(9) It is necessary to establish a number of new definitions in order to ensure that this Directive is uniformly applied by Member States. Where different obligations imposed by this Directive apply to different product categories and the relevant product falls into more than one of those categories (e.g. pipe, roll your-own tobacco), the stricter obligations should apply.

(10) Directive 2001/37/EC established maximum limits for tar, nicotine and carbon monoxide yields of cigarettes that should also be applicable to cigarettes which are exported from the Union. Those maximum limits and that approach remain valid.

(11) For measuring the tar, nicotine and carbon monoxide yields of cigarettes (hereinafter referred to as 'emission levels'), reference should be made to the relevant, internationally recognised ISO standards. The verification process should be protected from tobacco industry influence by using independent laboratories, including State laboratories. Member States should be able to use laboratories situated in other Member States of the Union. For other emissions from tobacco products, there are no internationally agreed standards or tests for quantifying maximum levels. The ongoing efforts at international level to develop such standards or tests should be encouraged.

(12) As regards establishing maximum emission levels, it could be necessary and appropriate at a later date to reduce the emission levels for tar, nicotine and carbon monoxide or to establish maximum levels for other emissions from tobacco products, taking into consideration their toxicity or addictiveness.

(13) In order to carry out their regulatory tasks, Member States and the Commission require comprehensive information on the ingredients and emissions from tobacco products to assess the attractiveness, addictiveness and toxicity of tobacco products and the health risks associated with the consumption of such products. To this end, the existing reporting obligations for ingredients and emissions should be strengthened. Additional enhanced reporting obligations should be provided for in respect of additives included in a priority list in order to assess, inter alia their toxicity, addictiveness and carcinogenic, mutagenic or reprotoxic properties ('CMR properties'), including in combusted form. The burden of such enhanced reporting obligations for SMEs should be limited to the extent possible. Such reporting obligations are consistent with the obligation placed on the Union to ensure a high level of protection for human health.

(14) The use of differing reporting formats, as is currently the case, makes it difficult for manufacturers and importers to fulfil their reporting obligations and burdensome for the Member States and the Commission to compare, analyse and draw conclusions from the information received. Therefore, there should be a common mandatory format for the reporting of ingredients and emissions. The greatest possible transparency of product information should be ensured

for the general public, whilst ensuring that appropriate account is taken of the trade secrets of the manufacturers of tobacco products. Existing systems for the reporting of ingredients should be taken into account.

(15) The lack of a harmonised approach to regulating the ingredients of tobacco products affects the smooth functioning of the internal market and has a negative impact on the free movement of goods across the Union. Some Member States have adopted legislation or entered into binding agreements with the industry allowing or prohibiting certain ingredients. As a result, some ingredients are regulated in certain Member States, but not in others. Member States also take differing approaches as regards additives in the filters of cigarettes as well as additives colouring the tobacco smoke. Without harmonisation, the obstacles to the smooth functioning of the internal market are expected to increase in the coming years, taking into account the implementation of the FCTC and the relevant FCTC guidelines throughout the Union and in the light of experience gained in other jurisdictions outside the Union. The FCTC guidelines in relation to the regulation of the contents of tobacco products and regulation of tobacco product disclosures call in particular for the removal of ingredients that increase palatability, create the impression that tobacco products have health benefits, are associated with energy and vitality or have colouring properties.

(16) The likelihood of diverging regulation is further increased by concerns over tobacco products having a characterising flavour other than one of tobacco, which could facilitate initiation of tobacco consumption or affect consumption patterns. Measures introducing unjustified differences of treatment between different types of flavoured cigarettes should be avoided. However, products with characterising flavour with a higher sales volume should be phased out over an extended time period to allow consumers adequate time to switch to other products.

(17) The prohibition of tobacco products with characterising flavours does not preclude the use of individual additives outright, but it does oblige manufacturers to reduce the additive or the combination of additives to such an extent that the additives no longer result in a characterising flavour. The use of additives necessary for the manufacture of tobacco products, for example sugar to replace sugar that is lost during the curing process, should be allowed, as long as they do not result in a characterising flavour or increase the addictiveness, toxicity or CMR properties of the product. An independent European advisory panel should assist in such decision making. The application of this Directive should not

lead to discrimination between different tobacco varieties, nor should it prevent product differentiation.

(18) Certain additives are used to create the impression that tobacco products have health benefits, present reduced health risks or increase mental alertness and physical performance. These additives, as well as additives that have CMR properties in unburnt form, should be prohibited in order to ensure uniform rules throughout the Union and a high level of protection of human health. Additives that increase addictiveness and toxicity should also be prohibited.

(19) Considering this Directive's focus on young people, tobacco products other than cigarettes and roll-your-own tobacco, should be granted an exemption from certain requirements relating to ingredients as long as there is no substantial change of circumstances in terms of sales volumes or consumption patterns of young people.

(20) Given the general prohibition of the sale of tobacco for oral use in the Union, the responsibility for regulating the ingredients of tobacco for oral use, which requires in-depth knowledge of the specific characteristics of this product and of its patterns of consumption, should, in accordance with the principle of subsidiarity, remain with Sweden, where the sale of this product is permitted pursuant to Article 151 of the Act of Accession of Austria, Finland and Sweden.

(21) In line with the purposes of this Directive, namely to facilitate the smooth functioning of the internal market for tobacco and related products, taking as a base a high level of health protection, especially for young people, and in line with Council Recommendation 2003/54/EC(1), Member States should be encouraged to prevent sales of such products to children and adolescents, by adopting appropriate measures that lay down and enforce age limits.

(22) Disparities still exist between national provisions regarding the labelling of tobacco products, in particular with regard to the use of combined health warnings consisting of a picture and a text, information on cessation services and promotional elements in and on unit packets.

(23) Such disparities are liable to constitute a barrier to trade and to impede the smooth functioning of the internal market in tobacco products, and should, therefore, be

---

(1) Council Recommendation 2003/54/EC of 2 December 2002 on the prevention of smoking and on initiatives to improve tobacco control (OJ L 22, 25.1.2003, p. 31).

eliminated. Also, it is possible that consumers in some Member States are better informed about the health risks of tobacco products than consumers in other Member States. Without further action at Union level, the existing disparities are likely to increase in the coming years.

(24) Adaptation of the provisions on labelling is also necessary to align the rules that apply at Union level to international developments. For example, the FCTC guidelines on the packaging and labelling of tobacco products call for large picture warnings on both principal display areas, mandatory cessation information and strict rules on misleading information. The provisions on misleading information will complement the general ban on misleading business to consumer commercial practices laid down in Directive 2005/29/EC of the European Parliament and of the Council(1).

Member States that use tax stamps or national identification marks for fiscal purposes on the packaging of tobacco products may, in some cases, have to provide for these stamps and marks to be repositioned in order to allow for the combined health warnings to be at the top of the principal display areas, in line with this Directive and the FCTC guidelines. Transitional arrangements should be put in place to allow Member States to maintain tax stamps or national identification marks used for fiscal purposes at the top of unit packets for a certain period after transposition of this Directive.

(25) The labelling provisions should also be adapted to new scientific evidence. For example, the indication of the emission levels for tar, nicotine and carbon monoxide on unit packets of cigarettes has proven to be misleading as it leads consumers to believe that certain cigarettes are less harmful than others. Evidence also suggests that large combined health warnings comprised of a text warning and a corresponding colour photograph are more effective than warnings consisting only of text. As a consequence, combined health warnings should become mandatory throughout the Union and cover significant and visible parts of the surface of unit packets. Minimum dimensions should be set for all health

---

(1) Directive 2005/29/EC of the European Parliament and of the Council of 11 May 2005 concerning unfair business-to-consumer commercial practices in the internal market and amending Council Directive 84/450/EEC, Directives 97/7/EC, 98/27/EC and 2002/65/EC of the European Parliament and of the Council and Regulation (EC) No 2006/2004 of the European Parliament and of the Council ('Unfair Commercial Practices Directive') (OJ L 149, 11.6.2005, p. 22).

warnings to ensure their visibility and effectiveness.

(26) For tobacco products for smoking, other than cigarettes and roll-your-own tobacco products, which are mainly consumed by older consumers and small groups of the population, it should be possible to continue to grant an exemption from certain labelling requirements as long as there is no substantial change of circumstances in terms of sales volumes or consumption patterns of young people. The labelling of these other tobacco products should follow rules that are specific to them. The visibility of health warnings on smokeless tobacco products should be ensured. Health warnings should, therefore, be placed on the two main surfaces of the packaging of smokeless tobacco products. As regards waterpipe tobacco, which is often perceived as less harmful than traditional tobacco products for smoking, the full labelling regime should apply in order to avoid consumers being misled.

(27) Tobacco products or their packaging could mislead consumers, in particular young people, where they suggest that these products are less harmful. This is, for example, the case if certain words or features are used, such as the words 'low-tar', 'light', 'ultra-light', 'mild', 'natural', 'organic', 'without additives', 'without flavours' or 'slim', or certain names, pictures, and figurative or other signs. Other misleading elements might include, but are not limited to, inserts or other additional material such as adhesive labels, stickers, onserts, scratch-offs and sleeves or relate to the shape of the tobacco product itself. Certain packaging and tobacco products could also mislead consumers by suggesting benefits in terms of weight loss, sex appeal, social status, social life or qualities such as femininity, masculinity or elegance. Likewise, the size and appearance of individual cigarettes could mislead consumers by creating the impression that they are less harmful. Neither the unit packets of tobacco products nor their outside packaging should include printed vouchers, discount offers, reference to free distribution, two-for-one or other similar offers that could suggest economic advantages to consumers thereby inciting them to buy those tobacco products.

(28) In order to ensure the integrity and the visibility of health warnings and maximise their efficacy, provisions should be made regarding the dimensions of the health warnings as well as regarding certain aspects of the appearance of the unit packets of tobacco products, including the shape and opening mechanism. When prescribing a cuboid shape for a unit packet, rounded or bevelled edges should be considered acceptable, provided the health warning covers a surface area that is equivalent to that on a unit packet without such edges. Member States apply

different rules on the minimum number of cigarettes per unit packet. Those rules should be aligned in order to ensure free circulation of the products concerned.

(29) Considerable volumes of illicit products, which do not fulfil the requirements laid down in Directive 2001/37/EC, are placed on the market and there are indications that these volumes might increase. Such illicit products undermine the free circulation of compliant products and the protection provided for by tobacco control legislation. In addition, the FCTC requires the Union to combat illicit tobacco products, including those illegally imported into the Union, as part of a comprehensive Union policy on tobacco control. Provision should, therefore, be made for unit packets of tobacco products to be marked with a unique identifier and security features and for their movements to be recorded so that such products can be tracked and traced throughout the Union and their compliance with this Directive can be monitored and better enforced. In addition, provision should be made for the introduction of security features that will facilitate the verification of whether or not tobacco products are authentic.

(30) An interoperable tracking and tracing system and security features should be developed at Union level. For an initial period only cigarettes and roll-your-own tobacco should be subjected to the tracking and tracing system and the security features. This would allow manufacturers of other tobacco products to benefit from the experience gained prior to the tracking and tracing system and security features becoming applicable to those other products.

(31) In order to ensure independence and transparency of the tracking and tracing system, manufacturers of tobacco products should conclude data storage contracts with independent third parties. The Commission should approve the suitability of those independent third parties and an independent external auditor should monitor their activities. The data related to the tracking and tracing system should be kept separate from other company related data and should be under the control of, and accessible at all times by, the competent authorities from Member States and the Commission.

(32) Council Directive 89/622/EEC[(1)] prohibited the sale in the Member States of

---

(1) Council Directive 89/622/EEC of 13 November 1989 on the approximation of the laws, regulations and administrative provisions of the Member States concerning the labelling of tobacco products and the prohibition of the marketing of certain types of tobacco for oral use (OJ L 359, 8.12.1989, p. 1).

certain types of tobacco for oral use. Directive 2001/37/EC reaffirmed that prohibition. Article 151 of the Act of Accession of Austria, Finland and Sweden grants Sweden a derogation from the prohibition. The prohibition of the sale of tobacco for oral use should be maintained in order to prevent the introduction in the Union (apart from Sweden) of a product that is addictive and has adverse health effects. For other smokeless tobacco products that are not produced for the mass market, strict provisions on labelling and certain provisions relating to their ingredients are considered sufficient to contain their expansion in the market beyond their traditional use.

(33) Cross-border distance sales of tobacco products could facilitate access to tobacco products that do not comply with this Directive. There is also an increased risk that young people would get access to tobacco products. Consequently, there is a risk that tobacco control legislation would be undermined. Member States should, therefore, be allowed to prohibit cross-border distance sales. Where cross-border distance sales are not prohibited, common rules on the registration of retail outlets engaging in such sales are appropriate to ensure the effectiveness of this Directive. Member States should, in accordance with Article 4(3) of the Treaty on European Union (TEU) cooperate with each other in order to facilitate the implementation of this Directive, in particular with respect to measures taken as regards cross-border distance sales of tobacco products.

(34) All tobacco products have the potential to cause mortality, morbidity and disability. Accordingly, their manufacture, distribution and consumption should be regulated. It is, therefore, important to monitor developments as regards novel tobacco products. Manufacturers and importers should be obliged to submit a notification of novel tobacco products, without prejudice to the power of the Member States to ban or to authorise such novel products.

(35) In order to ensure a level playing field, novel tobacco products, that are tobacco products as defined in this Directive, should comply with the requirements of this Directive.

(36) Electronic cigarettes and refill containers should be regulated by this Directive, unless they are - due to their presentation or function - subject to Directive 2001/83/EC of the European Parliament and of the Council[(1)] or to Council

---

(1) Directive 2001/83/EC of the European Parliament and of the Council of 6 November 2001 on the Community code relating to medicinal products for human use (OJ L 311, 28.11.2001, p. 67).

Directive 93/42/EEC[1]. Diverging legislation and practices as regards these products, including on safety requirements, exist between Member States, hence, action at Union level is required to improve the smooth functioning of the internal market. A high level of public health protection should be taken into account when regulating these products. In order to enable Member States to carry out their surveillance and control tasks, manufacturers and importers of electronic cigarettes and refill containers should be required to submit a notification of the relevant products before they are placed on the market.

(37) Member States should ensure that electronic cigarettes and refill containers comply with the requirements of this Directive. Where the manufacturer of the relevant product is not established in the Union, the importer of that product should bear the responsibilities relating to the compliance of those products with this Directive.

(38) Nicotine-containing liquid should only be allowed to be placed on the market under this Directive, where the nicotine concentration does not exceed 20 mg/ml. This concentration allows for a delivery of nicotine that is comparable to the permitted dose of nicotine derived from a standard cigarette during the time needed to smoke such a cigarette. In order to limit the risks associated with nicotine, maximum sizes for refill containers, tanks and cartridges should be set.

(39) Only electronic cigarettes that deliver nicotine doses at consistent levels should be allowed to be placed on the market under this Directive. Delivery of nicotine doses at consistent levels under normal conditions of use is necessary for health protection, safety and quality purposes, including to avoid the risk of accidental consumption of high doses.

(40) Electronic cigarettes and refill containers could create a health risk when in the hands of children. Therefore, it is necessary to ensure that such products are child- and tamperproof, including by means of child-proof labelling, fastenings and opening mechanisms.

(41) In view of the fact that nicotine is a toxic substance and considering the potential health and safety risks, including to persons for whom the product is not intended, nicotine-containing liquid should only be placed on the market in electronic cigarettes or in refill containers that meet certain safety and quality requirements.

---

(1) Council Directive 93/42/EEC of 14 June 1993 concerning medical devices (OJ L 169, 12.7.1993, p. 1).

It is important to ensure that electronic cigarettes do not break or leak during use and refill.

(42) The labelling and packaging of these products should display sufficient and appropriate information on their safe use, in order to protect human health and safety, should carry appropriate health warnings and should not include any misleading elements or features.

(43) Disparities between national laws and practices on advertising and sponsorship concerning electronic cigarettes present an obstacle to the free movement of goods and the freedom to provide services and create an appreciable risk of distortion of competition. Without further action at Union level, those disparities are likely to increase over the coming years, also taking into account the growing market for electronic cigarettes and refill containers. Therefore, it is necessary to approximate the national provisions on advertising and sponsorship of those products having cross-border effects, taking as a base a high level of protection of human health. Electronic cigarettes can develop into a gateway to nicotine addiction and ultimately traditional tobacco consumption, as they mimic and normalize the action of smoking. For this reason, it is appropriate to adopt a restrictive approach to advertising electronic cigarettes and refill containers.

(44) In order to perform their regulatory tasks, the Commission and Member States need comprehensive information on market developments as regards electronic cigarettes and refill containers. To this end manufacturers and importers of these products should be subject to reporting obligations on sales volumes, preference of various consumer groups and mode of sales. It should be ensured that this information is made available to the general public, taking the need to protect trade secrets duly into account.

(45) In order to ensure appropriate market surveillance by Member States, it is necessary that manufacturers, importers and distributors operate an appropriate system for monitoring and recording suspected adverse effects and inform the competent authorities about such effects so that appropriate action can be taken. It is warranted to provide for a safeguard clause that would allow Member States to take action to address serious risks to public health.

(46) In the context of an emerging market for electronic cigarettes, it is possible that, although complying with this Directive, specific electronic cigarettes or refill containers, or a type of electronic cigarette or refill container, placed on the market could pose an unforeseen risk to human health. It is therefore advisable to

provide for a procedure to address this risk, which should include the possibility for a Member State to adopt provisional appropriate measures. Such provisional appropriate measures could involve the prohibition of the placing on the market of specific electronic cigarettes or refill containers, or of a type of electronic cigarette or refill container. In this context, the Commission should be empowered to adopt delegated acts in order to prohibit the placing on the market of specific electronic cigarettes or refill containers, or of a type of electronic cigarette or refill container. The Commission should be empowered to do so, when at least three Member States have prohibited the products concerned on duly justified grounds and it is necessary to extend this prohibition to all Member States in order to ensure the smooth functioning of the internal market for products complying with this Directive but not presenting the same health risks. The Commission should report on the potential risks associated with refillable electronic cigarettes by 20 May 2016.

(47) This Directive does not harmonise all aspects of electronic cigarettes or refill containers. For example, the responsibility for adopting rules on flavours remains with the Member States. It could be useful for Member States to consider allowing the placing on the market of flavoured products. In doing so, they should be mindful of the potential attractiveness of such products for young people and non smokers. Any prohibition of such flavoured products would need to be justified and notification thereof submitted in accordance with Directive 98/34/EC of the European Parliament and of the Council[1].

(48) Moreover, this Directive does not harmonise the rules on smoke-free environments, or on domestic sales arrangements or domestic advertising, or brand stretching, nor does it introduce an age limit for electronic cigarettes or refill containers. In any case, the presentation and advertising of those products should not lead to the promotion of tobacco consumption or give rise to confusion with tobacco products. Member States are free to regulate such matters within the remit of their own jurisdiction and are encouraged to do so.

(49) The regulation of herbal products for smoking differs between Member States and these products are often perceived as harmless or less harmful despite the health risk caused by their combustion. In many cases consumers do not know

---

(1) Directive 98/34/EC of the European Parliament and of the Council of 22 June 1998 laying down a procedure for the provision of information in the field of technical standards and regulations and of rules on Information Society services (OJ L 204, 21.7.1998, p. 37).

the content of these products. In order to ensure the smooth functioning of the internal market and improve information to consumers, common labelling rules and ingredients reporting for these products should be introduced at Union level.

(50) In order to ensure uniform conditions for the implementation of this Directive implementing powers should be conferred on the Commission concerning the laying down and updating of a priority list of additives for enhanced reporting, the laying down and updating of the format for the reporting of ingredients and for the dissemination of that information, determining whether a tobacco product has a characterising flavour or has increased levels of toxicity, addictiveness or CMR properties, the methodology for determining whether a tobacco product has a characterising flavour, the procedures for the establishment and operation of an independent advisory panel for determining tobacco products with characterising flavours, the precise position of health warnings on pouches of roll-your-own tobacco, the technical specifications for the layout, design, and shape of combined health warnings, the technical standards for the establishment and operation of the tracking and tracing system, for ensuring the compatibility of the systems for the unique identifiers and for the security features, as well as establishing a common format for notification of electronic cigarettes and refill containers and the technical standards for the refill mechanisms for such products. Those implementing powers should be exercised in accordance with Regulation (EU) No 182/2011 of the European Parliament and of the Council[(1)].

(51) In order to ensure that this Directive is fully operational and to adapt it to technical, scientific and international developments in tobacco manufacture, consumption and regulation, the power to adopt acts in accordance with Article 290 TFEU should be delegated to the Commission in respect of adopting and adapting maximum emission levels and methods for measuring those emissions, setting maximum levels for additives that result in a characterising flavour or that increase toxicity or addictiveness, withdrawing certain exemptions granted to tobacco products other than cigarettes and roll-your-own tobacco, adapting the health warnings, establishing and adapting the picture library, defining the key elements of the data storage contracts to be concluded for the purposes of the tracking and tracing system, and extending measures adopted by Member States

---

(1) Regulation (EU) No 182/2011 of the European Parliament and of the Council of 16 February 2011 laying down the rules and general principles concerning mechanisms for control by Member States of the Commission's exercise of implementing powers (OJ L 55, 28.2.2011, p. 13).

to the entire Union concerning specific electronic cigarettes or refill containers or a type of electronic cigarette or refill container. It is of particular importance that the Commission carry out appropriate consultations during its preparatory work, including at expert level. The Commission, when preparing and drawing up delegated acts, should ensure a simultaneous, timely and appropriate transmission of relevant documents to the European Parliament and to the Council.

(52) The Commission should monitor the developments as regards the implementation and impact of this Directive and submit a report by 21 May 2021, and when necessary thereafter, in order to assess whether amendments to this Directive are necessary. The report should include information on the surfaces of unit packets of tobacco products that are not governed by this Directive, market developments concerning novel tobacco products, market developments that amount to a substantial change of circumstances, market developments concerning, and the consumer perception of, slim cigarettes, of waterpipe tobacco and of electronic cigarettes and refill containers.

The Commission should prepare a report regarding the feasibility, benefits and impact of a European system for the regulation of ingredients in tobacco products, including the feasibility and benefits of establishing a list of ingredients at Union level that can be used, or present in or added to tobacco products (so-called 'positive list'). In preparing that report, the Commission should evaluate, inter alia, the available scientific evidence on the toxic and addictive effects of ingredients.

(53) Tobacco and related products which comply with this Directive should benefit from the free movement of goods. However, in light of the different degrees of harmonisation achieved by this Directive, the Member States should, under certain conditions, retain the power to impose further requirements in certain respects in order to protect public health. This is the case in relation to the presentation and the packaging, including colours, of tobacco products other than health warnings, for which this Directive provides a first set of basic common rules. Accordingly, Member States could, for example, introduce provisions providing for further standardisation of the packaging of tobacco products, provided that those provisions are compatible with the TFEU, with WTO obligations and do not affect the full application of this Directive.

(54) Moreover, in order to take into account possible future market developments, Member States should also be allowed to prohibit a certain category of tobacco or related products, on grounds relating to the specific situation in the Member

State concerned and provided the provisions are justified by the need to protect public health, taking into account the high level of protection achieved through this Directive. Member States should notify such stricter national provisions to the Commission.

(55) A Member State should remain free to maintain or introduce national laws applying to all products placed on its national market for aspects not regulated by this Directive, provided they are compatible with the TFEU and do not jeopardise the full application of this Directive. Accordingly and under those conditions, a Member State could, inter alia, regulate or ban paraphernalia used for tobacco products (including waterpipes) and for herbal products for smoking as well as regulate or ban products resembling in appearance a type of tobacco or related product. Prior notification is required for national technical regulations pursuant to Directive 98/34/EC.

(56) Member States should ensure that personal data are only processed in accordance with the rules and safeguards laid down in Directive 95/46/EC of the European Parliament and of the Council(1).

(57) This Directive is without prejudice to Union laws governing the use and labelling of genetically modified organisms.

(58) In accordance with the Joint Political Declaration of 28 September 2011 of Member States and the Commission on explanatory documents(2), Member States have undertaken to accompany, in justified cases, the notification of their transposition measures with one or more documents explaining the relationship between the components of a directive and the corresponding parts of national transposition instruments. With regard to this Directive, the legislator considers the transmission of such documents to be justified.

(59) The obligation to respect the fundamental rights and legal principles enshrined in the Charter of Fundamental Rights of the European Union is not changed by this Directive. Several fundamental rights are affected by this Directive. It is therefore necessary to ensure that the obligations imposed on manufacturers, importers and distributors of tobacco and related products not only guarantee a high level

---

(1) Directive 95/46/EC of the European Parliament and of the Council of 24 October 1995 on the protection of individuals with regard to the processing of personal data and on the free movement of such data (OJ L 281, 23.11.1995, p. 31).

(2) OJ C 369, 17.12.2011, p. 14.

of health and consumer protection, but also protect all other fundamental rights and are proportionate with respect to the smooth functioning of the internal market. The application of this Directive should respect Union law and relevant international obligations.

(60) Since the objectives of this Directive, namely to approximate the laws, regulations and administrative provisions of the Member States concerning the manufacture, presentation and sale of tobacco and related products, cannot be sufficiently achieved by the Member States, but can rather, by reason of their scale and effects, be better achieved at Union level, the Union may adopt measures, in accordance with the principle of subsidiarity as set out in Article 5 TEU. In accordance with the principle of proportionality, as set out in that Article, this Directive does not go beyond what is necessary in order to achieve those objectives,

HAVE ADOPTED THIS DIRECTIVE:

TITLE I

COMMON PROVISIONS

Article 1

Subject matter

The objective of this Directive is to approximate the laws, regulations and administrative provisions of the Member States concerning:

(a) the ingredients and emissions of tobacco products and related reporting obligations, including the maximum emission levels for tar, nicotine and carbon monoxide for cigarettes;

(b) certain aspects of the labelling and packaging of tobacco products including the health warnings to appear on unit packets of tobacco products and any outside packaging as well as traceability and security features that are applied to tobacco products to ensure their compliance with this Directive;

(c) the prohibition on the placing on the market of tobacco for oral use;

(d) cross-border distance sales of tobacco products;

(e) the obligation to submit a notification of novel tobacco products;

(f) the placing on the market and the labelling of certain products, which are related to tobacco products, namely electronic cigarettes and refill containers, and herbal products for smoking;

in order to facilitate the smooth functioning of the internal market for tobacco and related products, taking as a base a high level of protection of human health, especially for young people, and to meet the obligations of the Union under the WHO Framework Convention for Tobacco Control ('FCTC').

Article 2

Definitions

For the purposes of this Directive, the following definitions shall apply:

(1) 'tobacco' means leaves and other natural processed or unprocessed parts of tobacco plants, including expanded and reconstituted tobacco;

(2) 'pipe tobacco' means tobacco that can be consumed via a combustion process and exclusively intended for use in a pipe;

(3) 'roll-your-own tobacco' means tobacco which can be used for making cigarettes by consumers or retail outlets;

(4) 'tobacco products' means products that can be consumed and consist, even partly, of tobacco, whether genetically modified or not;

(5) 'smokeless tobacco product' means a tobacco product not involving a combustion process, including chewing tobacco, nasal tobacco and tobacco for oral use;

(6) 'chewing tobacco' means a smokeless tobacco product exclusively intended for the purpose of chewing;

(7) 'nasal tobacco' means a smokeless tobacco product that can be consumed via the nose;

(8) 'tobacco for oral use' means all tobacco products for oral use, except those intended to be inhaled or chewed, made wholly or partly of tobacco, in powder or in particulate form or in any combination of those forms, particularly those presented in sachet portions or porous sachets;

(9) 'tobacco products for smoking' means tobacco products other than a smokeless tobacco product;

(10) 'cigarette' means a roll of tobacco that can be consumed via a combustion process and is further defined in Article 3(1) of Council Directive 2011/64/EU[1];

(11) 'cigar' means a roll of tobacco that can be consumed via a combustion process and is further defined in Article 4(1) of Directive 2011/64/EU;

(12) 'cigarillo' means a small type of cigar and is further defined in Article 8(1) of Council Directive 2007/74/EC[2];

(13) 'waterpipe tobacco' means a tobacco product that can be consumed via a waterpipe. For the purpose of this Directive, waterpipe tobacco is deemed to be a tobacco product for smoking. If a product can be used both via waterpipes and as roll-your-own tobacco, it shall be deemed to be roll-your-own tobacco;

(14) 'novel tobacco product' means a tobacco product which:

(a) does not fall into any of the following categories: cigarettes, roll-your-own tobacco, pipe tobacco, waterpipe tobacco, cigars, cigarillos, chewing tobacco, nasal tobacco or tobacco for oral use; and

(b) is placed on the market after 19 May 2014;

(15) 'herbal product for smoking' means a product based on plants, herbs or fruits which contains no tobacco and that can be consumed via a combustion process;

(16) 'electronic cigarette' means a product that can be used for consumption of nicotine-containing vapour via a mouth piece, or any component of that product, including a cartridge, a tank and the device without cartridge or tank. Electronic cigarettes can be disposable or refillable by means of a refill container and a tank, or rechargeable with single use cartridges;

(17) 'refill container' means a receptacle that contains a nicotine-containing liquid,

---

(1) Council Directive 2011/64/EU of 21 June 2011 on the structure and rates of excise duty applied to manufactured tobacco (OJ L 176, 5.7.2011, p. 24).

(2) Council Directive 2007/74/EC of 20 December 2007 on the exemption from value added tax and excise duty of goods imported by persons travelling from third countries (OJ L 346, 29.12.2007, p. 6).

which can be used to refill an electronic cigarette;

(18) 'ingredient' means tobacco, an additive, as well as any substance or element present in a finished tobacco product or related products, including paper, filter, ink, capsules and adhesives;

(19) 'nicotine' means nicotinic alkaloids;

(20) 'tar' means the raw anhydrous nicotine-free condensate of smoke;

(21) 'emissions' means substances that are released when a tobacco or related product is consumed as intended, such as substances found in smoke, or substances released during the process of using smokeless tobacco products;

(22) 'maximum level' or 'maximum emission level' means the maximum content or emission, including zero, of a substance in a tobacco product measured in milligrams;

(23) 'additive' means a substance, other than tobacco, that is added to a tobacco product, a unit packet or to any outside packaging;

(24) 'flavouring' means an additive that imparts smell and/or taste;

(25) 'characterising flavour' means a clearly noticeable smell or taste other than one of tobacco, resulting from an additive or a combination of additives, including, but not limited to, fruit, spice, herbs, alcohol, candy, menthol or vanilla, which is noticeable before or during the consumption of the tobacco product;

(26) 'addictiveness' means the pharmacological potential of a substance to cause addiction, a state which affects an individual's ability to control his or her behaviour, typically by instilling a reward or a relief from withdrawal symptoms, or both;

(27) 'toxicity' means the degree to which a substance can cause harmful effects in the human organism, including effects occurring over time, usually through repeated or continuous consumption or exposure;

(28) 'substantial change of circumstances' means an increase of the sales volumes by product category by at least 10 % in at least five Member States based on sales data transmitted in accordance with Article 5(6) or an increase of the level of prevalence of use in the under 25 years of age consumer group by at least five percentage points in at least five Member States for the respective

product category based on the Special Eurobarometer 385 report of May 2012 or equivalent prevalence studies; in any case, a substantial change of circumstances is deemed not to have occurred if the sales volume of the product category at retail level does not exceed 2,5 % of total sales of tobacco products at Union level;

(29) 'outside packaging' means any packaging in which tobacco or related products are placed on the market and which includes a unit packet or an aggregation of unit packets; transparent wrappers are not regarded as outside packaging;

(30) 'unit packet' means the smallest individual packaging of a tobacco or related product that is placed on the market;

(31) 'pouch' means a unit packet of roll-your own tobacco, either in the form of a rectangular pocket with a flap that covers the opening or in the form of a standing pouch;

(32) 'health warning' means a warning concerning the adverse effects on human health of a product or other undesired consequences of its consumption, including text warnings, combined health warnings, general warnings and information messages, as provided for in this Directive;

(33) 'combined health warning' means a health warning consisting of a combination of a text warning and a corresponding photograph or illustration, as provided for in this Directive;

(34) 'cross-border distance sales' means distance sales to consumers where, at the time the consumer orders the product from a retail outlet, the consumer is located in a Member State other than the Member State or the third country where that retail outlet is established; a retail outlet is deemed to be established in a Member State:

(a) in the case of a natural person: if he or she has his or her place of business in that Member State;

(b) in other cases: if the retail outlet has its statutory seat, central administration or place of business, including a branch, agency or any other establishment, in that Member State;

(35) 'consumer' means a natural person who is acting for purposes which are outside his or her trade, business, craft or profession;

(36) 'age verification system' means a computing system that unambiguously confirms

the consumer's age electronically in accordance with national requirements;

(37) 'manufacturer' means any natural or legal person who manufactures a product or has a product designed or manufactured, and markets that product under their name or trademark;

(38) 'import of tobacco or related products' means the entry into the territory of the Union of such products unless the products are placed under a customs suspensive procedure or arrangement upon their entry into the Union, as well as their release from a customs suspensive procedure or arrangement;

(39) 'importer of tobacco or related products' means the owner of, or a person having the right of disposal over, tobacco or related products that have been brought into the territory of the Union;

(40) 'placing on the market' means to make products, irrespective of their place of manufacture, available to consumers located in the Union, with or without payment, including by means of distance sale; in the case of cross-border distance sales the product is deemed to be placed on the market in the Member State where the consumer is located;

(41) 'retail outlet' means any outlet where tobacco products are placed on the market including by a natural person.

TITLE II

TOBACCO PRODUCTS

CHAPTER I

Ingredients and emissions

Article 3

Maximum emission levels for tar, nicotine, carbon monoxide and other substances

1. The emission levels from cigarettes placed on the market or manufactured in the Member States ('maximum emission levels') shall not be greater than:

(a) 10 mg of tar per cigarette;

(b) 1 mg of nicotine per cigarette;

(c) 10 mg of carbon monoxide per cigarette.

2. The Commission shall be empowered to adopt delegated acts in accordance with Article 27 to decrease the maximum emission levels laid down in paragraph 1, where this is necessary based on internationally agreed standards.

3. Member States shall notify the Commission of any maximum emission levels they set for emissions from cigarettes other than the emissions referred to in paragraph 1 and for emissions from tobacco products other than cigarettes.

4. The Commission shall adopt delegated acts in accordance with Article 27 to integrate standards agreed by the parties to the FCTC or by the WHO relating to maximum emission levels for emissions from cigarettes other than the emissions referred to in paragraph 1 and for emissions from tobacco products other than cigarettes into Union law.

Article 4

Measurement methods

1. The tar, nicotine and carbon monoxide emissions from cigarettes shall be measured on the basis of ISO standard 4387 for tar, ISO standard 10315 for nicotine, and ISO standard 8454 for carbon monoxide.

The accuracy of the tar, nicotine and carbon monoxide measurements shall be determined in accordance with ISO standard 8243.

2. The measurements referred to in paragraph 1 shall be verified by laboratories which are approved and monitored by the competent authorities of the Member States.

Those laboratories shall not be owned or controlled directly or indirectly by the tobacco industry.

Member States shall communicate to the Commission a list of approved laboratories, specifying the criteria used for approval and the methods of monitoring applied, and shall update that list whenever any change is made. The Commission shall make those lists of approved laboratories publicly available.

3. The Commission shall be empowered to adopt delegated acts in accordance with

Article 27 to adapt the methods of measurement of the tar, nicotine and carbon monoxide emissions, where this is necessary, based on scientific and technical developments or internationally agreed standards.

4. Member States shall notify the Commission of any measurement methods they use for emissions from cigarettes other than the emissions referred to in paragraph 3 and for emissions from tobacco products other than cigarettes.

5. The Commission shall adopt delegated acts in accordance with Article 27 to integrate standards agreed by the parties to the FCTC or by the WHO for measurement methods into Union law.

6. Member States may charge manufacturers and importers of tobacco products proportionate fees for the verification of the measurements referred to in paragraph 1 of this Article.

Article 5

Reporting of ingredients and emissions

1. Member States shall require manufacturers and importers of tobacco products to submit to their competent authorities the following information by brand name and type:

(a) a list of all ingredients, and quantities thereof, used in the manufacture of the tobacco products, in descending order of the weight of each ingredient included in the tobacco products;

(b) the emission levels referred to in Article 3(1) and (4);

(c) where available, information on other emissions and their levels.

For products already placed on the market that information shall be provided by 20 November 2016.

Manufacturers or importers shall also inform the competent authorities of the Member States concerned, if the composition of a product is modified in a way that affects the information provided under this Article.

For a new or modified tobacco product the information required under this Article shall be submitted prior to the placing on the market of those products.

2. The list of ingredients referred to in point (a) of paragraph 1 shall be accompanied by a statement setting out the reasons for the inclusion of such ingredients in the tobacco products concerned. That list shall also indicate the status of the ingredients, including whether they have been registered under Regulation (EC) No 1907/2006 of the European Parliament and of the Council[(1)] as well as their classification under Regulation (EC) No 1272/2008 of the European Parliament and of the Council[(2)].

3. The list referred to in point (a) of paragraph 1 shall also be accompanied by the relevant toxicological data regarding the ingredients in burnt or unburnt form, as appropriate, referring in particular to their effects on the health of consumers and taking into account, inter alia, any addictive effects.

Furthermore, for cigarettes and roll-your-own tobacco, a technical document setting out a general description of the additives used and their properties, shall be submitted by the manufacturer or importer.

Other than for tar, nicotine and carbon monoxide and for emissions referred to in Article 4(4), manufacturers and importers shall indicate the methods of measurement of emissions used. Member States may also require manufacturers or importers to carry out studies as may be prescribed by the competent authorities in order to assess the effects of ingredients on health, taking into account, inter alia, their addictiveness and toxicity.

4. Member States shall ensure that the information submitted in accordance with paragraph 1 of this Article and of Article 6 is made publicly available on a website. The Member States shall take the need to protect trade secrets duly into account when making that information publicly available. Member States shall require manufacturers and importers to specify, when submitting the information pursuant to paragraph 1

---

(1) Regulation (EC) No 1907/2006 of the European Parliament and of the Council of 18 December 2006 concerning the Registration, Evaluation, Authorisation and Restriction of Chemicals (REACH), establishing a European Chemicals Agency, amending Directive 1999/45/EC and repealing Council Regulation (EEC) No 793/93 and Commission Regulation (EC) No 1488/94 as well as Council Directive 76/769/EEC and Commission Directives 91/155/EEC, 93/67/EEC, 93/105/EC and 2000/21/EC (OJ L 396, 30.12.2006, p. 1).

(2) Regulation (EC) No 1272/2008 of the European Parliament and of the Council of 16 December 2008 on classification, labelling and packaging of substances and mixtures, amending and repealing Directives 67/548/EEC and 1999/45/EC, and amending Regulation (EC) No 1907/2006 (OJ L 353, 31.12.2008, p. 1).

of this Article and Article 6, the information which they consider to constitute trade secrets.

5. The Commission shall, by means of implementing acts, lay down and, if necessary, update the format for the submission and the making available of information referred to in paragraphs 1 and 6 of this Article and Article 6. Those implementing acts shall be adopted in accordance with the examination procedure referred to in Article 25(2).

6. Member States shall require manufacturers and importers to submit internal and external studies available to them on market research and preferences of various consumer groups, including young people and current smokers, relating to ingredients and emissions, as well as executive summaries of any market surveys they carry out when launching new products. Member States shall also require manufacturers and importers to report their sales volumes per brand and type, reported in sticks or kilograms, and per Member State on a yearly basis starting from 1 January 2015. Member States shall provide any other sales volume data that is available to them.

7. All data and information to be provided to and by Member States under this Article and under Article 6 shall be provided in electronic form. Member States shall store the information electronically and shall ensure that the Commission and other Member States have access to that information for the purposes of applying this Directive. Member States and the Commission shall ensure that trade secrets and other confidential information are treated in a confidential manner.

8. Member States may charge manufacturers and importers of tobacco products proportionate fees for receiving, storing, handling, analysing and publishing the information submitted to them pursuant to this Article.

Article 6

Priority list of additives and enhanced reporting obligations

1. In addition to the reporting obligations laid down in Article 5, enhanced reporting obligations shall apply to certain additives contained in cigarettes and roll-your-own tobacco that are included in a priority list. The Commission shall adopt implementing acts laying down and subsequently updating such a priority list of additives. This list shall contain additives:

(a) for which initial indications, research, or regulation in other jurisdictions exist suggesting that they have one of the properties set out in points (a) to (d) of paragraph 2 of this Article; and

(b) which are amongst the most commonly used additives by weight or number according to the reporting of ingredients pursuant to paragraphs 1 and 3 of Article 5.

Those implementing acts shall be adopted in accordance with the examination procedure referred to in Article 25(2). A first list of additives shall be adopted by 20 May 2016 and shall contain at least 15 additives.

2. Member States shall require manufacturers and importers of cigarettes and roll-your-own tobacco containing an additive that is included in the priority list provided for in paragraph 1, to carry out comprehensive studies, which shall examine for each additive whether it:

(a) contributes to the toxicity or addictiveness of the products concerned, and whether this has the effect of increasing the toxicity or addictiveness of any of the products concerned to a significant or measurable degree;

(b) results in a characterising flavour;

(c) facilitates inhalation or nicotine uptake; or

(d) leads to the formation of substances that have CMR properties, the quantities thereof, and whether this has the effect of increasing the CMR properties in any of the products concerned to a significant or measurable degree.

3. Those studies shall take into account the intended use of the products concerned and examine in particular the emissions resulting from the combustion process involving the additive concerned. The studies shall also examine the interaction of that additive with other ingredients contained in the products concerned. Manufacturers or importers using the same additive in their tobacco products may carry out a joint study when using that additive in a comparable product composition.

4. Manufacturers or importers shall establish a report on the results of these studies. That report shall include an executive summary, and a comprehensive overview compiling the available scientific literature on that additive and summarising internal data on the effects of the additive.

Manufacturers or importers shall submit these reports to the Commission and a copy thereof to the competent authorities of those Member States where a tobacco product containing this additive is placed on the market at the latest 18 months after the additive concerned has been included in the priority list pursuant to paragraph 1.

The Commission and the Member States concerned may also request supplementary information from manufacturers or importers regarding the additive concerned. This supplementary information shall form part of the report.

The Commission and the Member States concerned may require these reports to be peer reviewed by an independent scientific body, in particular as regards their comprehensiveness, methodology and conclusions. The information received shall assist the Commission and Member States in taking the decisions pursuant to Article 7. The Member States and the Commission may charge manufacturers and importers of tobacco products proportionate fees for those peer reviews.

5. Small and medium-sized enterprises as defined in Commission Recommendation 2003/361/EC[(1)] shall be exempted from the obligations pursuant to this Article, if a report on that additive is prepared by another manufacturer or importer.

Article 7

Regulation of ingredients

1. Member States shall prohibit the placing on the market of tobacco products with a characterising flavour.

Member States shall not prohibit the use of additives which are essential for the manufacture of tobacco products, for example sugar to replace sugar that is lost during the curing process, provided those additives do not result in a product with a characterising flavour and do not increase to a significant or measureable degree the addictiveness, toxicity or the CMR properties of the tobacco product.

Member States shall notify the Commission of the measures taken pursuant to this paragraph.

2. The Commission shall, at the request of a Member State, or may, on its own initiative, determine by means of implementing acts whether a tobacco product falls within the scope of paragraph 1. Those implementing acts shall be adopted in accordance with the examination procedure referred to in Article 25(2).

3. The Commission shall adopt implementing acts laying down uniform rules for

(1) Commission Recommendation 2003/361/EC of 6 May 2003 concerning the definition of micro, small and medium-sized enterprises (OJ L 124, 20.5.2003, p. 36).

the procedures for determining whether a tobacco product falls within the scope of paragraph 1. Those implementing acts shall be adopted in accordance with the examination procedure referred to in Article 25(2).

4. An independent advisory panel shall be established at Union level. Member States and the Commission may consult this panel before adopting a measure pursuant to paragraphs 1 and 2 of this Article. The Commission shall adopt implementing acts laying down the procedures for the establishment and operation of this panel.

Those implementing acts shall be adopted in accordance with the examination procedure referred to in Article 25(2).

5. Where the content level or concentration of certain additives or the combination thereof has resulted in prohibitions pursuant to paragraph 1 of this Article in at least three Member States, the Commission shall be empowered to adopt delegated acts in accordance with Article 27 to set maximum content levels for those additives or combination of additives that result in the characterising flavour.

6. Member States shall prohibit the placing on the market of tobacco products containing the following additives:

(a) vitamins or other additives that create the impression that a tobacco product has a health benefit or presents reduced health risks;

(b) caffeine or taurine or other additives and stimulant compounds that are associated with energy and vitality;

(c) additives having colouring properties for emissions;

(d) for tobacco products for smoking, additives that facilitate inhalation or nicotine uptake; and

(e) additives that have CMR properties in unburnt form.

7. Member States shall prohibit the placing on the market of tobacco products containing flavourings in any of their components such as filters, papers, packages, capsules or any technical features allowing modification of the smell or taste of the tobacco products concerned or their smoke intensity. Filters, papers and capsules shall not contain tobacco or nicotine.

8. Member States shall ensure that the provisions and conditions laid down in Regulation (EC) No 1907/2006 are applied to tobacco products as appropriate.

9. Member States shall, on the basis of scientific evidence, prohibit the placing on the market of tobacco products containing additives in quantities that increase the toxic or addictive effect, or the CMR properties of a tobacco product at the stage of consumption to a significant or measureable degree.

Member States shall notify to the Commission the measures they have taken pursuant to this paragraph.

10. The Commission shall, at the request of a Member State, or may, on its own initiative, determine by means of an implementing act whether a tobacco product falls within the scope of paragraph 9. Those implementing acts shall be adopted in accordance with the examination procedure referred to in Article 25(2) and shall be based on the latest scientific evidence.

11. Where an additive or a certain quantity thereof has been shown to amplify the toxic or addictive effect of a tobacco product, and where this has resulted in prohibitions pursuant to paragraph (9) of this Article in at least three Member States, the Commission shall be empowered to adopt delegated acts in accordance with Article 27 to set maximum content levels for those additives. In this case, the maximum content level shall be set at the lowest maximum level that led to one of the national prohibitions referred to in this paragraph.

12. Tobacco products other than cigarettes and roll-your-own tobacco shall be exempted from the prohibitions laid down in paragraphs 1 and 7. The Commission shall adopt delegated acts in accordance with Article 27 to withdraw that exemption for a particular product category, if there is a substantial change of circumstances as established in a Commission report.

13. The Member States and the Commission may charge proportionate fees to manufacturers and importers of tobacco products for assessing whether a tobacco product has a characterising flavour, whether prohibited additives or flavourings are used and whether a tobacco product contains additives in quantities that increase to a significant and measurable degree the toxic or addictive effect or the CMR properties of the tobacco product concerned.

14. In the case of tobacco products with a characterising flavour whose Union-wide sales volumes represent 3 % or more in a particular product category, the provisions of this Article shall apply from 20 May 2020.

15. This Article shall not apply to tobacco for oral use.

CHAPTER II

Labelling and packaging

Article 8

General provisions

1. Each unit packet of a tobacco product and any outside packaging shall carry the health warnings provided for in this Chapter in the official language or languages of the Member State where the product is placed on the market.

2. Health warnings shall cover the entire surface of the unit packet or outside packaging that is reserved for them and they shall not be commented on, paraphrased or referred to in any form.

3. Member States shall ensure that the health warnings on a unit packet and any outside packaging are irremovably printed, indelible and fully visible, including not being partially or totally hidden or interrupted by tax stamps, price marks, security features, wrappers, jackets, boxes, or other items, when tobacco products are placed on the market. On unit packets of tobacco products other than cigarettes and roll-your-own tobacco in pouches, the health warnings may be affixed by means of stickers, provided that such stickers are irremovable. The health warnings shall remain intact when opening the unit packet other than packets with a flip-top lid, where the health warnings may be split when opening the packet, but only in a manner that ensures the graphical integrity and visibility of the text, photographs and cessation information.

4. The health warnings shall in no way hide or interrupt the tax stamps, price marks, tracking and tracing marks, or security features on unit packets.

5. The dimensions of the health warnings provided for in Articles 9, 10, 11 and 12 shall be calculated in relation to the surface concerned when the packet is closed.

6. Health warnings shall be surrounded by a black border of a width of 1 mm inside the surface area that is reserved for these warnings, except for health warnings pursuant to Article 11.

7. When adapting a health warning pursuant to Articles 9(5), 10(3) and 12(3), the Commission shall ensure that it is factual or that Member States shall have a choice of two warnings, one of which is factual.

8. Images of unit packets and any outside packaging targeting consumers in the Union

shall comply with the provisions of this chapter.

Article 9

General warnings and information messages on tobacco products for smoking

1. Each unit packet and any outside packaging of tobacco products for smoking shall carry one of the following general warnings:

'Smoking kills – quit now'

or

'Smoking kills'

Member States shall determine which of the general warnings referred to in the first subparagraph is to be used.

2. Each unit packet and any outside packaging of tobacco products for smoking shall carry the following information message:

'Tobacco smoke contains over 70 substances known to cause cancer.'

3. For cigarette packets and roll-your-own tobacco in cuboid packets the general warning shall appear on the bottom part of one of the lateral surfaces of the unit packets, and the information message shall appear on the bottom part of the other lateral surface. These health warnings shall have a width of not less than 20 mm.

For packets in the form of a shoulder box with a hinged lid that result in the lateral surfaces being split into two when the packet is open, the general warning and the information message shall appear in their entirety on the larger parts of those split surfaces. The general warning shall also appear on the inside of the top surface that is visible when the packet is open.

The lateral surfaces of this type of packet shall have a height of not less than 16 mm.

For roll-your-own tobacco marketed in pouches the general warning and the information message shall appear on the surfaces that ensure the full visibility of those health warnings. For roll-your-own tobacco in cylindrical packets the general warning shall appear on the outside surface of the lid and the information message on the inside surface of the lid.

Both the general warning and the information message shall cover 50 % of the surfaces on which they are printed.

4. The general warning and information message referred to in paragraphs 1 and 2 shall be:

(a) printed in black Helvetica bold type on a white background. In order to accommodate language requirements, Member States may determine the font size, provided that the font size specified in national law ensures that the relevant text occupies the greatest possible proportion of the surface reserved for these health warnings; and

(b) at the centre of the surface reserved for them, and on cuboid packets and any outside packaging they shall be parallel to the lateral edge of the unit packet or of the outside packaging.

5. The Commission shall be empowered to adopt delegated acts in accordance with Article 27 to adapt the wording of the information message laid down in paragraph 2 to scientific and market developments.

6. The Commission shall, by means of implementing acts, determine the precise position of the general warning and the information message on roll-your-own tobacco marketed in pouches, taking into account the different shapes of pouches.

Those implementing acts shall be adopted in accordance with the examination procedure referred to in Article 25(2).

Article 10

Combined health warnings for tobacco products for smoking

1. Each unit packet and any outside packaging of tobacco products for smoking shall carry combined health warnings. The combined health warnings shall:

(a) contain one of the text warnings listed in Annex I and a corresponding colour photograph specified in the picture library in Annex II;

(b) include smoking cessation information such as telephone numbers, e-mail addresses or Internet sites intending to inform consumers about the programmes that are available to support persons who want to stop smoking;

(c) cover 65 % of both the external front and back surface of the unit packet and any outside packaging. Cylindrical packets shall display two combined health warnings, equidistant from each other, each covering 65 % of their respective half of the curved surface;

(d) show the same text warning and corresponding colour photograph on both sides of the unit packets and any outside packaging;

(e) appear at the top edge of a unit packet and any outside packaging, and be positioned in the same direction as any other information appearing on that surface of the packaging. Transitional exemptions from that obligation on the position of the combined health warning may apply in Member States where tax stamps or national identification marks used for fiscal purposes remain mandatory, as follows:

(i) in those cases, where the tax stamp or national identification mark used for fiscal purposes is affixed at the top edge of a unit packet made of carton material, the combined health warning that is to appear on the back surface may be positioned directly below the tax stamp or national identification mark;

(ii) where a unit packet is made of soft material, Member States may allow for a rectangular area to be reserved for the tax stamp or national identification mark used for fiscal purposes of a height not exceeding 13 mm between the top edge of the packet and the top end of the combined health warnings.

The exemptions referred to in points (i) and (ii) shall apply for a period of three years from 20 May 2016. Brand names or logos shall not be positioned above the health warnings;

(f) be reproduced in accordance with the format, layout, design and proportions specified by the Commission pursuant to paragraph 3;

(g) in the case of unit packets of cigarettes, respect the following dimensions:

(i) height: not less than 44 mm;

(ii) width: not less than 52 mm.

2. The combined health warnings are grouped into three sets as set out in Annex II and each set shall be used in a given year and rotated on an annual basis. Member States shall ensure that each combined health warning available for use in a given year is displayed to the extent possible in equal numbers on each brand of tobacco products.

3. The Commission shall be empowered to adopt delegated acts in accordance with Article 27 to:

(a) adapt the text warnings listed in Annex I taking into account scientific and market developments;

(b) establish and adapt the picture library referred to in point (a) of paragraph 1 of this Article taking into account scientific and market developments.

4. The Commission shall by means of implementing acts define the technical specifications for the layout, design and shape of the combined health warnings, taking into account the different packet shapes.

Those implementing acts shall be adopted in accordance with the examination procedure referred to in Article 25(2).

Article 11

Labelling of tobacco products for smoking other than cigarettes, roll-your-own tobacco and waterpipe tobacco

1. Member States may exempt tobacco products for smoking other than cigarettes, roll-your-own tobacco and waterpipe tobacco from the obligations to carry the information message laid down in Article 9(2) and the combined health warnings laid down in Article 10. In that event, and in addition to the general warning provided for in Article 9(1), each unit packet and any outside packaging of such products shall carry one of the text warnings listed in Annex I. The general warning specified in Article 9(1) shall include a reference to the cessation services referred to in Article 10(1)(b).

The general warning shall appear on the most visible surface of the unit packet and any outside packaging.

Member States shall ensure that each text warning is displayed to the extent possible in equal numbers on each brand of these products. The text warnings shall appear on the next most visible surface of the unit packet and any outside packaging.

For unit packets with a hinged lid, the next most visible surface is the one that becomes visible when the packet is open.

2. The general warning referred to in paragraph 1 shall cover 30 % of the relevant

surface of the unit packet and any outside packaging. That proportion shall be increased to 32 % for Member States with two official languages and to 35 % for Member States with more than two official languages.

3. The text warning referred to in paragraph 1 shall cover 40 % of the relevant surface of the unit packet and any outside packaging. That proportion shall be increased to 45 % for Member States with two official languages and 50 % for Member States with more than two official languages.

4. Where the health warnings referred to in paragraph 1 are to appear on a surface exceeding 150 cm^2 , the warnings shall cover an area of 45 cm^2 . That area shall be increased to 48 cm^2 for Member States with two official languages and 52,5 cm^2 for Member States with more than two official languages.

5. The health warnings referred to in paragraph 1 shall comply with the requirements specified in Article 9(4). The text of the health warnings shall be parallel to the main text on the surface reserved for these warnings.

The health warnings shall be surrounded by a black border of a width of not less than 3 mm and not more than 4 mm. This border shall appear outside the surface reserved for the health warnings.

6. The Commission shall adopt delegated acts in accordance with Article 27, to withdraw the possibility of granting exemptions for any of the particular product categories referred to in paragraph 1 if there is a substantial change of circumstances as established in a Commission report for the product category concerned.

Article 12

Labelling of smokeless tobacco products

1. Each unit packet and any outside packaging of smokeless tobacco products shall carry the following health warning:

'This tobacco product damages your health and is addictive.'

2. The health warning laid down in paragraph 1 shall comply with the requirements specified in Article 9(4). The text of the health warnings shall be parallel to the main text on the surface reserved for these warnings.

In addition, it shall:

(a) appear on the two largest surfaces of the unit packet and any outside packaging;

(b) cover 30 % of the surfaces of the unit packet and any outside packaging. That proportion shall be increased to 32 % for Member States with two official languages and 35 % for Member States with more than two official languages.

3. The Commission shall be empowered to adopt delegated acts in accordance with Article 27 to adapt the wording of the health warning laid down in paragraph 1 to scientific developments.

Article 13

Product presentation

1. The labelling of unit packets and any outside packaging and the tobacco product itself shall not include any element or feature that:

(a) promotes a tobacco product or encourages its consumption by creating an erroneous impression about its characteristics, health effects, risks or emissions; labels shall not include any information about the nicotine, tar or carbon monoxide content of the tobacco product;

(b) suggests that a particular tobacco product is less harmful than others or aims to reduce the effect of some harmful components of smoke or has vitalising, energetic, healing, rejuvenating, natural, organic properties or has other health or lifestyle benefits;

(c) refers to taste, smell, any flavourings or other additives or the absence thereof;

(d) resembles a food or a cosmetic product;

(e) suggests that a certain tobacco product has improved biodegradability or other environmental advantages.

2. The unit packets and any outside packaging shall not suggest economic advantages by including printed vouchers, offering discounts, free distribution, two-for-one or other similar offers.

3. The elements and features that are prohibited pursuant to paragraphs 1 and 2 may include but are not limited to texts, symbols, names, trademarks, figurative or other signs.

Article 14

Appearance and content of unit packets

1. Unit packets of cigarettes shall have a cuboid shape. Unit packets of roll-your-own tobacco shall have a cuboid or cylindrical shape, or the form of a pouch. A unit packet of cigarettes shall include at least 20 cigarettes. A unit packet of roll-your-own tobacco shall contain tobacco weighing not less than 30 g.

2. A unit packet of cigarettes may consist of carton or soft material and shall not have an opening that can be reclosed or re-sealed after it is first opened, other than the flip-top lid and shoulder box with a hinged lid. For packets with a flip-top lid and hinged lid, the lid shall be hinged only at the back of the unit packet.

Article 15

Traceability

1. Member States shall ensure that all unit packets of tobacco products are marked with a unique identifier. In order to ensure the integrity of the unique identifier, it shall be irremovably printed or affixed, indelible and not hidden or interrupted in any form, including through tax stamps or price marks, or by the opening of the unit packet. In the case of tobacco products that are manufactured outside of the Union, the obligations laid down in this Article apply only to those that are destined for, or placed on, the Union market.

2. The unique identifier shall allow the following to be determined:

(a) the date and place of manufacturing;

(b) the manufacturing facility;

(c) the machine used to manufacture the tobacco products;

(d) the production shift or time of manufacture;

(e) the product description;

(f) the intended market of retail sale;

(g) the intended shipment route;

(h) where applicable, the importer into the Union;

(i) the actual shipment route from manufacturing to the first retail outlet, including all warehouses used as well as the shipment date, shipment destination, point of departure and consignee;

(j) the identity of all purchasers from manufacturing to the first retail outlet; and

(k) the invoice, order number and payment records of all purchasers from manufacturing to the first retail outlet.

3. The information referred to in points (a), (b), (c), (d), (e), (f), (g) and, where applicable, (h) of paragraph 2 shall form part of the unique identifier.

4. Member States shall ensure that the information mentioned in points (i), (j) and (k) of paragraph 2 is electronically accessible by means of a link to the unique identifier.

5. Member States shall ensure that all economic operators involved in the trade of tobacco products, from the manufacturer to the last economic operator before the first retail outlet, record the entry of all unit packets into their possession, as well as all intermediate movements and the final exit of the unit packets from their possession. This obligation may be complied with by the marking and recording of aggregated packaging such as cartons, mastercases or pallets, provided that the tracking and tracing of all unit packets remains possible.

6. Member States shall ensure that all natural and legal persons engaged in the supply chain of tobacco products maintain complete and accurate records of all relevant transactions.

7. Member States shall ensure that the manufacturers of tobacco products provide all economic operators involved in the trade of tobacco products, from the manufacturer to the last economic operator before the first retail outlet, including importers, warehouses and transporting companies, with the equipment that is necessary for the recording of the tobacco products purchased, sold, stored, transported or otherwise handled. That equipment shall be able to read and transmit the recorded data electronically to a data storage facility pursuant to paragraph 8.

8. Member States shall ensure that manufacturers and importers of tobacco products conclude data storage contracts with an independent third party, for the purpose of hosting the data storage facility for all relevant data. The data storage facility shall be physically located on the territory of the Union. The suitability of the third party,

in particular its independence and technical capacities, as well as the data storage contract, shall be approved by the Commission.

The third party's activities shall be monitored by an external auditor, who is proposed and paid by the tobacco manufacturer and approved by the Commission. The external auditor shall submit an annual report to the competent authorities and to the Commission, assessing in particular any irregularities in relation to access.

Member States shall ensure that the Commission, the competent authorities of the Member States, and the external auditor have full access to the data storage facilities. In duly justified cases the Commission or the Member States may grant manufacturers or importers access to the stored data, provided that commercially sensitive information remains adequately protected in conformity with the relevant Union and national law.

9. Recorded data shall not be modified or deleted by an economic operator involved in the trade of tobacco products.

10. Member States shall ensure that personal data are only processed in accordance with the rules and safeguards laid down in Directive 95/46/EC.

11. The Commission shall, by means of implementing acts:

(a) determine the technical standards for the establishment and the operation of the tracking and tracing system as provided for in this Article, including the marking with a unique identifier, the recording, transmitting, processing and storing of data and access to stored data;

(b) determine the technical standards for ensuring that the systems used for the unique identifier and the related functions are fully compatible with each other across the Union.

Those implementing acts shall be adopted in accordance with the examination procedure referred to in Article 25(2).

12. The Commission shall be empowered to adopt delegated acts in accordance with Article 27 to define the key elements of the data storage contracts referred to in paragraph 8 of this Article, such as duration, renewability, expertise required or confidentiality, including the regular monitoring and evaluation of those contracts.

13. Paragraphs 1 to 10 shall apply to cigarettes and roll-your-own tobacco from 20

May 2019 and to tobacco products other than cigarettes and roll-your-own tobacco from 20 May 2024.

Article 16

Security feature

1. In addition to the unique identifier referred to in Article 15, Member States shall require that all unit packets of tobacco products, which are placed on the market, carry a tamper proof security feature, composed of visible and invisible elements. The security feature shall be irremovably printed or affixed, indelible and not hidden or interrupted in any form, including through tax stamps and price marks, or other elements imposed by legislation.

Member States requiring tax stamps or national identification marks used for fiscal purposes may allow that they are used for the security feature provided that the tax stamps or national identification marks fulfil all of the technical standards and functions required under this Article.

2. The Commission shall, by means of implementing acts, define the technical standards for the security feature and their possible rotation and adapt them to scientific, market and technical developments.

Those implementing acts shall be adopted in accordance with the examination procedure referred to in Article 25(2).

3. Paragraph 1 shall apply to cigarettes and roll-your-own tobacco from 20 May 2019 and to tobacco products other than cigarettes and roll-your-own tobacco from 20 May 2024.

CHAPTER III

Tobacco for oral use, cross-border distance sales of tobacco products and novel tobacco products

Article 17

Tobacco for oral use

Member States shall prohibit the placing on the market of tobacco for oral use, without prejudice to Article 151 of the Act of Accession of Austria, Finland and Sweden.

Article 18

Cross-border distance sales of tobacco products

1. Member States may prohibit cross-border distance sales of tobacco products to consumers. Member States shall cooperate to prevent such sales. Retail outlets engaging in cross-border distance sales of tobacco products may not supply such products to consumers in Member States where such sales have been prohibited. Member States which do not prohibit such sales shall require retail outlets intending to engage in cross-border distance sales to consumers located in the Union to register with the competent authorities in the Member State, where the retail outlet is established, and in the Member State, where the actual or potential consumers are located. Retail outlets established outside the Union shall be required to register with the competent authorities in the Member State where the actual or potential consumers are located. All retail outlets intending to engage in cross-border distance sales shall submit at least the following information to the competent authorities when registering:

(a) name or corporate name and permanent address of the place of activity from where the tobacco products will be supplied;

(b) the starting date of the activity of offering tobacco products for cross-border distance sales to consumers by means of Information Society services, as defined in point 2 of Article 1 of Directive 98/34/EC;

(c) the address of the website or websites used for that purpose and all relevant information necessary to identify the website.

2. The competent authorities of the Member States shall ensure that consumers have access to the list of all retail outlets registered with them. When making that list available, Member States shall ensure that the rules and safeguards laid down in Directive 95/46/EC are complied with. Retail outlets may only start placing tobacco products on the market via cross-border distance sales when they have received confirmation of their registration with the relevant competent authority.

3. The Member States of destination of tobacco products sold via cross-border distance sales may require that the supplying retail outlet nominates a natural person

to be responsible for verifying — before the tobacco products reach the consumer — that they comply with the national provisions adopted pursuant to this Directive in the Member State of destination, if such verification is necessary in order to ensure compliance and facilitate enforcement.

4. Retail outlets engaged in cross-border distance sales shall operate an age verification system, which verifies, at the time of sale, that the purchasing consumer complies with minimum age requirements provided for under the national law of the Member State of destination. The retail outlet or natural person nominated pursuant to paragraph 3 shall provide to the competent authorities of that Member State a description of the details and functioning of the age verification system.

5. Retail outlets shall only process personal data of the consumer in accordance with Directive 95/46/EC and those data shall not be disclosed to the manufacturer of tobacco products or companies forming part of the same group of companies or to other third parties. Personal data shall not be used or transferred for purposes other than the actual purchase. This also applies if the retail outlet forms part of a manufacturer of tobacco products.

Article 19

Notification of novel tobacco products

1. Member Stes shall require manufacturers and importers of novel tobacco products to submit a notification to the competent authorities of Member States of any such product they intend to place on the national market concerned. The notification shall be submitted in electronic form six months before the intended placing on the market. It shall be accompanied by a detailed description of the novel tobacco product concerned as well as instructions for its use and information on ingredients and emissions in accordance with Article 5. The manufacturers and importers submitting a notification of a novel tobacco product shall also provide the competent authorities with:

(a) available scientific studies on toxicity, addictiveness and attractiveness of the novel tobacco product, in particular as regards its ingredients and emissions;

(b) available studies, executive summaries thereof and market research on the preferences of various consumer groups, including young people and current smokers;

(c) other available and relevant information, including a risk/benefit analysis of the product, its expected effects on cessation of tobacco consumption, its expected effects on initiation of tobacco consumption and predicted consumer perception.

2. Member States shall require manufacturers and importers of novel tobacco products to transmit to their competent authorities any new or updated information on the studies, research and other information referred to in points (a) to (c) of paragraph 1. Member States may require manufacturers or importers of novel tobacco products to carry out additional tests or submit additional information. Member States shall make all information received pursuant to this Article available to the Commission.

3. Member States may introduce a system for the authorisation of novel tobacco products. Member States may charge manufacturers and importers proportionate fees for that authorisation.

4. Novel tobacco products placed on the market shall respect the requirements of this Directive. Which of the provisions of this Directive apply to novel tobacco products depends on whether those products fall under the definition of a smokeless tobacco product or of a tobacco product for smoking.

TITLE III

ELECTRONIC CIGARETTES AND HERBAL PRODUCTS FOR SMOKING

Article 20

Electronic cigarettes

1. The Member States shall ensure that electronic cigarettes and refill containers are only placed on the market if they comply with this Directive and with all other relevant Union legislation.

This Directive does not apply to electronic cigarettes and refill containers that are subject to an authorisation requirement under Directive 2001/83/EC or to the requirements set out in Directive 93/42/EEC.

2. Manufacturers and importers of electronic cigarettes and refill containers shall submit a notification to the competent authorities of the Member States of any such products which they intend to place on the market. The notification shall be submitted in electronic form six months before the intended placing on the market. For electronic

cigarettes and refill containers already placed on the market on 20 May 2016, the notification shall be submitted within six months of that date. A new notification shall be submitted for each substantial modification of the product.

The notification shall, depending on whether the product is an electronic cigarette or a refill container, contain the following information:

(a) the name and contact details of the manufacturer, a responsible legal or natural person within the Union, and, if applicable, the importer into the Union;

(b) a list of all ingredients contained in, and emissions resulting from the use of, the product, by brand name and type, including quantities thereof;

(c) toxicological data regarding the product's ingredients and emissions, including when heated, referring in particular to their effects on the health of consumers when inhaled and taking into account, inter alia, any addictive effect;

(d) information on the nicotine doses and uptake when consumed under normal or reasonably foreseeable conditions;

(e) a description of the components of the product; including, where applicable, the opening and refill mechanism of the electronic cigarette or refill containers;

(f) a description of the production process, including whether it involves series production, and a declaration that the production process ensures conformity with the requirements of this Article;

(g) a declaration that the manufacturer and importer bear full responsibility for the quality and safety of the product, when placed on the market and used under normal or reasonably foreseeable conditions.

Where Member States consider that the information submitted is incomplete, they shall be entitled to request the completion of the information concerned.

Member States may charge manufacturers and importers proportionate fees for receiving, storing, handling and analysing the information submitted to them.

3. Member States shall ensure that:

(a) nicotine-containing liquid is only placed on the market in dedicated refill containers not exceeding a volume of 10 ml, in disposable electronic cigarettes or in single use cartridges and that the cartridges or tanks do not exceed a volume of 2 ml;

(b) the nicotine-containing liquid does not contain nicotine in excess of 20 mg/ml;

(c) the nicotine-containing liquid does not contain additives listed in Article 7(6);

(d) only ingredients of high purity are used in the manufacture of the nicotine-containing liquid. Substances other than the ingredients referred to in point (b) of the second subparagraph of paragraph 2 of this Article are only present in the nicotine-containing liquid in trace levels, if such traces are technically unavoidable during manufacture;

(e) except for nicotine, only ingredients are used in the nicotine-containing liquid that do not pose a risk to human health in heated or unheated form;

(f) electronic cigarettes deliver the nicotine doses at consistent levels under normal conditions of use;

(g) electronic cigarettes and refill containers are child- and tamper-proof, are protected against breakage and leakage and have a mechanism that ensures refilling without leakage.

4. Member States shall ensure that:

(a) unit packets of electronic cigarettes and refill containers include a leaflet with information on:

  (i) instructions for use and storage of the product, including a reference that the product is not recommended for use by young people and non-smokers;

  (ii) contra-indications;

  (iii) warnings for specific risk groups;

  (iv) possible adverse effects;

  (v) addictiveness and toxicity; and

  (vi) contact details of the manufacturer or importer and a legal or natural contact person within the Union;

(b) unit packets and any outside packaging of electronic cigarettes and refill containers:

  (i) include a list of all ingredients contained in the product in descending order of the weight, and an indication of the nicotine content of the product and the delivery per dose, the batch number and a recommendation to keep the product out of reach of children;

(ii) without prejudice to point (i) of this point, do not include elements or features referred to in Article 13, with the exception of Article 13(1)(a) and (c) concerning information on the nicotine content and on flavourings; and

(iii) carry one of the following health warnings:

'This product contains nicotine which is a highly addictive substance. It is not recommended for use by nonsmokers'.

or

'This product contains nicotine which is a highly addictive substance.'

Member States shall determine which of these health warnings is to be used;

(c) health warnings comply with the requirements specified in Article 12(2).

5. Member States shall ensure that:

(a) commercial communications in Information Society services, in the press and other printed publications, with the aim or direct or indirect effect of promoting electronic cigarettes and refill containers are prohibited, except for publications that are intended exclusively for professionals in the trade of electronic cigarettes or refill containers and for publications which are printed and published in third countries, where those publications are not principally intended for the Union market;

(b) commercial communications on the radio, with the aim or direct or indirect effect of promoting electronic cigarettes and refill containers, are prohibited;

(c) any form of public or private contribution to radio programmes with the aim or direct or indirect effect of promoting electronic cigarettes and refill containers is prohibited;

(d) any form of public or private contribution to any event, activity or individual person with the aim or direct or indirect effect of promoting electronic cigarettes and refill containers and involving or taking place in several Member States or otherwise having cross-border effects is prohibited;

(e) audiovisual commercial communications to which Directive 2010/13/EU of the European Parliament and of the Council[1] applies, are prohibited for electronic cigarettes and refill containers.

6. Article 18 of this Directive shall apply to cross-border distance sales of electronic cigarettes and refill containers.

7. Member States shall require manufacturers and importers of electronic cigarettes and refill containers to submit, annually, to the competent authorities:

(i) comprehensive data on sales volumes, by brand name and type of the product;

(ii) information on the preferences of various consumer groups, including young people, non-smokers and the main types of current users;

(iii) the mode of sale of the products; and

(iv) executive summaries of any market surveys carried out in respect of the above, including an English translation thereof.

Member States shall monitor the market developments concerning electronic cigarettes and refill containers, including any evidence that their use is a gateway to nicotine addiction and ultimately traditional tobacco consumption among young people and non-smokers.

8. Member States shall ensure that the information received pursuant to paragraph 2 is made publicly available on a website. The Member States shall take the need to protect trade secrets duly into account when making that information publicly available.

Member States shall, upon request, make all information received pursuant to this Article available to the Commission and other Member States. The Member States and the Commission shall ensure that trade secrets and other confidential information are treated in a confidential manner.

9. Member States shall require manufacturers, importers and distributers of electronic

(1) Directive 2010/13/EU of the European Parliament and of the Council of 10 March 2010 on the coordination of certain provisions laid down by law, regulation or administrative action in Member States concerning the provision of audiovisual media services (Audiovisual Media Services Directive) (OJ L 95, 15.4.2010, p. 1).

cigarettes and refill containers to establish and maintain a system for collecting information about all of the suspected adverse effects on human health of these products.

Should any of these economic operators consider or have reason to believe that electronic cigarettes or refill containers, which are in their possession and are intended to be placed on the market or are placed on the market, are not safe or are not of good quality or are otherwise not in conformity with this Directive, that economic operator shall immediately take the corrective action necessary to bring the product concerned into conformity with this Directive, to withdraw or to recall it, as appropriate. In such cases the economic operator shall also be required to immediately inform the market surveillance authorities of the Member States in which the product is made available or is intended to be made available, giving details, in particular, of the risk to human health and safety and of any corrective action taken, and of the results of such corrective action.

Member States may also request additional information from the economic operators, for example on the safety and quality aspects or any adverse effects of electronic cigarettes or refill containers.

10. The Commission shall submit a report to the European Parliament and the Council on the potential risks to public health associated with the use of refillable electronic cigarettes by 20 May 2016 and whenever appropriate thereafter.

11. In the case of electronic cigarettes and refill containers that comply with the requirements of this Article, where a competent authority ascertains or has reasonable grounds to believe that specific electronic cigarettes or refill containers, or a type of electronic cigarette or refill container, could present a serious risk to human health, it may take appropriate provisional measures. It shall immediately inform the Commission and the competent authorities of other Member States of the measures taken and shall communicate any supporting data. The Commission shall determine, as soon as possible after having received that information, whether the provisional measure is justified. The Commission shall inform the Member State concerned of its conclusions to enable the Member State to take appropriate follow-up measures.

Where, in application of the first subparagraph of this paragraph, the placing on the market of specific electronic cigarettes or refill containers, or a type of electronic cigarette or refill container has been prohibited on duly justified grounds in at least three Member States, the Commission shall be empowered to adopt delegated acts in accordance with Article 27 to extend such a prohibition to all Member States, if such an extension is justified and proportionate.

12. The Commission shall be empowered to adopt delegated acts in accordance with Article 27 to adapt the wording of the health warning in paragraph 4(b) of this Article. When adapting that health warning, the Commission shall ensure that it is factual.

13. The Commission shall, by means of an implementing act, lay down a common format for the notification provided for in paragraph 2 and technical standards for the refill mechanism provided for in paragraph 3(g).

These implementing acts shall be adopted in accordance with the examination procedure referred to in Article 25(2).

Article 21

Herbal products for smoking

1. Each unit packet and any outside packaging of herbal products for smoking shall carry the following health warning:

'Smoking this product damages your health.'

2. The health warning shall be printed on the front and back external surface of the unit packet and on any outside packaging.

3. The health warning shall comply with the requirements set out in Article 9(4). It shall cover 30 % of the area of the corresponding surface of the unit packet and of any outside packaging. That proportion shall be increased to 32 % for Member States with two official languages and to 35 % for Member States with more than two official languages.

4. Unit packets and any outside packaging of herbal products for smoking shall not include any of the elements or features set out in Article 13(1)(a), (b) and (d) and shall not state that the product is free of additives or flavourings.

Article 22

Reporting of ingredients of herbal products for smoking

1. Member States shall require manufacturers and importers of herbal products for

smoking to submit to their competent authorities a list of all ingredients, and quantities thereof that are used in the manufacture of such products by brand name and type. Manufacturers or importers shall also inform the competent authorities of the Member States concerned when the composition of a product is modified in a way that affects the information submitted pursuant to this Article. The information required under this Article shall be submitted prior to the placing on the market of a new or modified herbal product for smoking.

2. Member States shall ensure that the information submitted in accordance with paragraph 1 is made publicly available on a website. The Member States shall take the need to protect trade secrets duly into account when making that information publicly available. Economic operators shall specify exactly which information they consider to constitute a trade secret.

TITLE IV

FINAL PROVISIONS

Article 23

Cooperation and enforcement

1. Member States shall ensure that manufacturers and importers of tobacco and related products provide the Commission and the competent authorities of the Member States with complete and correct information requested pursuant to this Directive and within the time limits set out herein. The obligation to provide the requested information shall lie primarily with the manufacturer, if the manufacturer is established in the Union. The obligation to provide the requested information shall lie primarily with the importer, if the manufacturer is established outside the Union and the importer is established inside the Union. The obligation to provide the requested information shall lie jointly with the manufacturer and the importer if both are established outside the Union.

2. Member States shall ensure that tobacco and related products which do not comply with this Directive, including the implementing and delegated acts provided for therein, are not placed on the market. Member States shall ensure that tobacco and related products are not placed on the market if the reporting obligations set out in this Directive are not complied with.

3. Member States shall lay down rules on penalties applicable to infringements of the

national provisions adopted pursuant to this Directive and shall take all measures that are necessary to ensure that these penalties are enforced. The penalties provided for shall be effective, proportionate and dissuasive. Any financial administrative penalty that may be imposed as a result of an intentional infringement may be such as to offset the economic advantage sought through the infringement.

4. The competent authorities of the Member States shall cooperate with each other and with the Commission to ensure the correct application and due enforcement of this Directive and shall transmit to each other all information necessary with a view to applying this Directive in a uniform manner.

Article 24

Free movement

1. Member States may not, for considerations relating to aspects regulated by this Directive, and subject to paragraphs 2 and 3 of this Article, prohibit or restrict the placing on the market of tobacco or related products which comply with this Directive.

2. This Directive shall not affect the right of a Member State to maintain or introduce further requirements, applicable to all products placed on its market, in relation to the standardisation of the packaging of tobacco products, where it is justified on grounds of public health, taking into account the high level of protection of human health achieved through this Directive. Such measures shall be proportionate and may not constitute a means of arbitrary discrimination or a disguised restriction on trade between Member States. Those measures shall be notified to the Commission together with the grounds for maintaining or introducing them.

3. A Member State may also prohibit a certain category of tobacco or related products, on grounds relating to the specific situation in that Member State and provided the provisions are justified by the need to protect public health, taking into account the high level of protection of human health achieved through this Directive. Such national provisions shall be notified to the Commission together with the grounds for introducing them. The Commission shall, within six months of the date of receiving the notification provided for in this paragraph, approve or reject the national provisions after having verified, taking into account the high level of protection of human health achieved through this Directive, whether or not they are justified, necessary and proportionate to their aim and whether or not they are a means of

arbitrary discrimination or a disguised restriction on trade between the Member States. In the absence of a decision by the Commission within the period of six months, the national provisions shall be deemed to be approved.

Article 25

Committee procedure

1. The Commission shall be assisted by a committee. That committee shall be a committee within the meaning of Regulation (EU) No 182/2011.

2. Where reference is made to this paragraph, Article 5 of Regulation (EU) No 182/2011 shall apply.

3. Where the opinion of the committee is to be obtained by written procedure, that procedure shall be terminated without result when, within the time-limit for delivery of the opinion, the chair of the committee so decides or a simple majority of committee members so requests.

4. Where the Committee delivers no opinion, the Commission shall not adopt the draft implementing act and the third subparagraph of Article 5(4) of Regulation (EU) No 182/2011 shall apply.

Article 26

Competent authorities

Member States shall designate the competent authorities that shall be responsible for the implementation and enforcement of the obligations provided for in this Directive within three months of 20 May 2016. Member States shall inform the Commission about the identity of the designated authorities without delay. The Commission shall publish that information in the *Official Journal of the European Union*.

Article 27

Exercise of the delegation

1. The power to adopt delegated acts is conferred on the Commission subject to the conditions laid down in this Article.

2. The power to adopt delegated acts referred to in Articles 3(2) and (4), 4(3) and (5), 7(5), (11) and (12), 9(5), 10(3), 11(6), 12(3), 15(12), 20(11) and (12) shall be conferred on the Commission for a period of five years from 19 May 2014. The Commission shall draw up a report in respect of the delegation of power not later than nine months before the end of the five-year period. The delegation of power shall be tacitly extended for periods of an identical duration, unless the European Parliament or the Council opposes such extension not later than three months before the end of each period.

3. The delegation of powers referred to in Articles 3(2) and(4), 4(3) and (5), 7(5), (11) and (12), 9(5), 10(3), 11(6), 12(3), 15(12), 20(11) and (12) may be revoked at any time by the European Parliament or by the Council. A decision to revoke shall put an end to the delegation of the power specified in that decision. It shall take effect the day following the publication of the decision in the *Official Journal of the European Union* or at a later date specified therein. It shall not affect the validity of any delegated acts already in force.

4. As soon as it adopts a delegated act, the Commission shall notify it simultaneously to the European Parliament and to the Council.

5. A delegated act adopted pursuant to Articles 3(2) and (4), 4(3) and (5), 7(5), (11) and (12), 9(5), 10(3), 11(6), 12(3), 15(12), 20(11) and (12) shall enter into force only if no objection has been expressed either by the European Parliament or the Council within a period of two months of notification of that act to the European Parliament and the Council or if, before the expiry of that period, the European Parliament and the Council have both informed the Commission that they will not object. That period shall be extended by two months at the initiative of the European Parliament or of the Council.

Article 28

Report

1. No later than five years from 20 May 2016, and whenever necessary thereafter, the Commission shall submit to the European Parliament, the Council, the European Economic and Social Committee and the Committee of the Regions a report on the application of this Directive.

When drafting the report, the Commission shall be assisted by scientific and technical experts in order to have all the necessary information at its disposal.

2. In the report, the Commission shall indicate, in particular, the elements of the Directive which should be reviewed or adapted in the light of scientific and technical developments, including the development of internationally agreed rules and standards on tobacco and related products. The Commission shall pay special attention to:

(a) the experience gained with respect to the design of package surfaces not governed by this Directive taking into account national, international, legal, economic and scientific developments;

(b) market developments concerning novel tobacco products considering, inter alia, notifications received under Article 19;

(c) market developments which constitute a substantial change of circumstances;

(d) the feasibility, benefits and possible impact of a European system for the regulation of the ingredients used in tobacco products, including the establishment, at Union level, of a list of ingredients that may be used or present in, or added to tobacco products, taking into account, inter alia, the information collected in accordance with Articles 5 and 6;

(e) market developments concerning cigarettes with a diameter of less than 7,5 mm, and consumer perception of their harmfulness as well as the misleading character of such cigarettes;

(f) the feasibility, benefits and possible impact of a Union database containing information on ingredients and emissions from tobacco products collected in accordance with Articles 5 and 6;

(g) market developments concerning electronic cigarettes and refill containers considering, amongst others, information collected in accordance with Article 20, including on the initiation of consumption such products by young people and non-smokers and the impact of such products on cessation efforts as well as measures taken by Member States regarding flavours;

(h) market developments and consumer preferences as regards waterpipe tobacco, with a particular focus on its flavours.

The Member States shall assist the Commission and provide all available information for carrying out the assessment and preparing the report.

3. The report shall be followed-up by proposals for amending this Directive, which the Commission deem necessary to adapt it - to the extent necessary for the smooth

functioning of the internal market - to developments in the field of tobacco and related products, and to take into account new developments based on scientific facts and developments concerning internationally agreed standards for tobacco and related products.

Article 29

Transposition

1. Member States shall bring into force the laws, regulations and administrative provisions necessary to comply with this Directive by 20 May 2016. They shall forthwith communicate to the Commission the text of those provisions.

The Member States shall apply those measures from 20 May 2016, without prejudice to Articles 7(14), 10(1)(e), 15(13) and 16(3).

2. When Member States adopt these provisions, they shall contain a reference to this Directive or be accompanied by such reference on the occasion of their official publication. They shall also include a statement that references in existing laws, regulations and administrative provisions to the Directive repealed by this Directive shall be construed as references to this Directive. The Member States shall determine how such reference is to be made and how that statement is to be formulated.

3. Member States shall communicate to the Commission the text of the main provisions of national law which they adopt in the field covered by this Directive.

Article 30

Transitional provision

Member States may allow the following products, which are not in compliance with this Directive, to be placed on the market until 20 May 2017:

(a) tobacco products manufactured or released for free circulation and labelled in accordance with Directive 2001/37/EC before 20 May 2016;

(b) electronic cigarettes or refill containers manufactured or released for free circulation before 20 November 2016;

(c) herbal products for smoking manufactured or released for free circulation before

20 May 2016.

Article 31

Repeal

Directive 2001/37/EC is repealed with effect from 20 May 2016, without prejudice to the obligations of the Member States relating to the time-limits for the transposition into national law of that Directive.

References to the repealed Directive shall be construed as references to this Directive and read in accordance with the correlation table in Annex III to this Directive.

Article 32

Entry into force

This Directive shall enter into force on the twentieth day following that of its publication in the *Official Journal of the European Union*.

Article 33

Addressees

This Directive is addressed to the Member States.

Done at Brussels, 3 April 2014.

For the European Parliament

The President

M. SCHULZ

For the Council

The President

D. KOURKOULAS

ANNEX I

LIST OF TEXT WARNINGS

(referred to in Article 10 and Article 11(1))

(1) Smoking causes 9 out of 10 lung cancers

(2) Smoking causes mouth and throat cancer

(3) Smoking damages your lungs

(4) Smoking causes heart attacks

(5) Smoking causes strokes and disability

(6) Smoking clogs your arteries

(7) Smoking increases the risk of blindness

(8) Smoking damages your teeth and gums

(9) Smoking can kill your unborn child

(10) Your smoke harms your children, family and friends

(11) Smokers'children are more likely to start smoking

(12) Quit smoking – stay alive for those close to you

(13) Smoking reduces fertility

(14) Smoking increases the risk of impotence

ANNEX II

PICTURE LIBRARY

(REFERRED TO IN ARTICLE 10(1))

[To be established by the Commission pursuant to Article 10(3)(b).]

ANNEX III

CORRELATION TABLE

Directive 2001/37/EC	This Directive
Article 1	Article 1
Article 2	Article 2
Article 3(1)	Article 3(1)
Article 3(2) and (3)	—
Article 4(1)	Article 4(1)
Article 4(2)	Article 4(2)
Article 4(3) to (5)	—
Article 5(1)	—
Article 5(2) point (a)	Article 9(1)
Article 5(2) point (b)	Article 10(1) point (a) and 10(2), Article 11(1)
Article 5(3)	Article 10(1)
Article 5(4)	Article 12
Article 5(5) first subparagraph	Article 9(3) fifth subparagraph Article 11(2) and (3) Article 12(2) point (b)
Article 5(5) second subparagraph	Article 11(4)
Article 5(6) point (a)	Article 9(4) point (a)
Article 5(6) point (b)	—
Article 5(6) point (c)	Article 9(4) point(b)
Article 5(6) point (d)	Article 8(6) and Article 11(5) second subparagraph
Article 5(6) point (e)	Article 8(1)
Article 5(7)	Article 8(3) and (4)
Article 5(8)	—
Article 5(9) first subparagraph	Article 15(1) and (2)
Article 5(9) second subparagraph	Article 15(11)
Article 6 (1) first subparagraph	Article 5(1) first subparagraph
Article 6 (1) second subparagraph	Article 5(2) and (3)
Article 6 (1) third subparagraph	—
Article 6(2)	Article 5(4)
Article 6(3) and (4)	—
Article 7	Article 13(1) point (b)
Article 8	Article 17
Article 9(1)	Article 4(3)
Article 9(2)	Article 10(2) and (3) point (a)
Article 9(3)	Article 16(2)
Article 10(1)	Article 25(1)

Directive 2001/37/EC	This Directive
Article 10(2) and (3)	Article 25(2)
Article 11 first and second subparagraphs	Article 28(1) first and second subparagraphs
Article 11 third subparagraph	Article 28(2) first subparagraph
Article 11 fourth subparagraph	Article 28(3)
Article 12	—
Article 13(1)	Article 24(1)
Article 13(2)	Article 24(2)
Article 13(3)	
Article 14(1) first subparagraph	Article 29(1) first subparagraph
Article 14(1) second subparagraph	Article 29(2)
Article 14(2) and (3)	Article 30 point (a)
Article 14(4)	Article 29(3)
Article 15	Article 31
Article 16	Article 32
Article 17	Article 33
Annex I (List of additional health warnings)	Annex I (List of text warnings)
Annex II (Time-limits for transposition and implementation of repealed Directives)	—
Annex III (Correlation table)	Annex III (Correlation table)